International Yearbook of Soil Law and Policy

Regional Perspectives

Series Editor
Harald Ginzky, German Environment Agency, Dessau, Germany

IYSLP *Regional Perspectives* is a subseries of the International Yearbook of Soil Law and Policy (IYSLP). It discusses the sustainable management of soils, in particular from specific regional perspectives. Topics include concepts and models of soil protection law, tenure rights issues, and gender and cultural dimensions of soil protection policies in specific regions. The series also addresses the need for international cooperation, taking into account specific regional requirements and challenges. As such, IYSLP *Regional Perspectives* offers an indispensable information tool for policymakers, scientists and lawyers. All volumes are peer-reviewed by the series editor.

More information about this subseries at http://www.springer.com/series/16057

Hadijah Yahyah • Harald Ginzky •
Emmanuel Kasimbazi • Robert Kibugi •
Oliver C. Ruppel

Editors

Legal Instruments for Sustainable Soil Management in Africa

 Springer

Editors
Hadijah Yahyah
School of Law
Kampala International University
Kampala, Uganda

Harald Ginzky
German Environment Agency
Dessau, Germany

Development and Rule of Law
Programme [DROP]
University of Stellenbosch
Stellenbosch, South Africa

Emmanuel Kasimbazi
School of Law
Makerere University
Kampala, Uganda

Robert Kibugi
School of Law
University of Nairobi
Nairobi, Kenya

Oliver C. Ruppel
Development and Rule of Law
Programme [DROP]
University of Stellenbosch
Stellenbosch, South Africa

Fraunhofer Centre for International
Management and Knowledge Economy [IMW]
Leipzig, Germany

ISSN 2520-1271   ISSN 2520-128X   (electronic)
International Yearbook of Soil Law and Policy
ISSN 2524-4094   ISSN 2524-4108   (electronic)
Regional Perspectives
ISBN 978-3-030-36006-1   ISBN 978-3-030-36004-7   (eBook)
https://doi.org/10.1007/978-3-030-36004-7

This Springer imprint is published by the registered company Springer Nature Switzerland AG.
The registered company address is: Gewerbestrasse 11, 6330 Cham, Switzerland

Foreword

Towards Sustainable Management of Soils: The German–African Cooperation

The workshop entitled "Legal instruments for the effective protection and sustainable management of soils" was organized in cooperation with Germany and Uganda. Germany and Uganda are long-standing partners and can look back on more than 50 years of successful and trustful cooperation.

SDG 15—Life on Land—states as follows: Human life depends on the earth as much as the ocean for our sustenance and livelihoods. Plant life provides 80% of our human diet, and we rely on agriculture as an important economic resource and means of development. Forests account for 30% of the earth's surface, providing vital habitats for millions of species and important sources for clean air and water, as well as being crucial for combating climate change.

Today we are witnessing unprecedented land degradation; drought and desertification is also on the rise each year. Of the 8300 animal breeds known, 8% are extinct and a further 22% are at risk of extinction.

Land is one of the most essential pillars of human existence and economic development. In Uganda, land is the most invaluable asset for citizens.

It is acquired to provide livelihood and facilitate production and economic transformation of the country. Consequentially, land issues are highly sensitive in Uganda. In a bid to resolve the rampant conflicts over land in the country, a commission of inquiry into land matters was recently established.

Secured access to land is one of the key prerequisites for the sustainable use and protection of soils and, as such, is fundamental to poverty and hunger reduction. In this connection, Uganda has created a favorable framework for reform of land law through progressive constitutional and land policy. Four tenure regimes are available: freehold, leasehold, customary land, and Mailo land. However, fewer than 10% of all plots of land in Uganda are formally registered. Rural populations are particularly prone to lack the necessary documentation to protect their land rights. For this reason, traditional land ownership rights can often be established only with testimony from neighbors, village elders, or clan representatives. Issuing land titles or land certificates is a lengthy and costly process and, for most of the rural population, simply prohibitive.

The GIZ project "Responsible Land Policy in Uganda" is working in this area to ensure systematic documentation of the land rights of those living in rural areas. It strives for the land documentation of customary ownership rights of 5000 households in the Teso subregion and of 75,000 land parcels in Central Uganda on private Mailo land. Examples from other countries demonstrate a positive impact of land tenure on national security and food security: once farmers have rights to cultivate land for a longer period, they are inclined to make long-term investments into improved agriculture productivity and soil conservation.

The German response to food security challenges is the special initiative ONE WORLD – No Hunger, launched in early 2014. With annual funding of 1.5 billion Euros, we aim to fight hunger and malnutrition and to enable future generations to feed themselves. It is implemented in Benin, Madagascar, Laos, Peru, and Uganda. The aim of the program in Uganda is to achieve land documentation for 80,000 households and to reduce land-related conflicts.

This initiative also presents immense potential for upscaling, and we invite other donors and agencies to join forces with us.

In addition, Germany has enthusiastically supported the international Initiative on "Economics of Land Degradation (ELD)." In October 2015, the ELD Initiative together with UNEP published the report "The Economics of Land Degradation in Africa," focusing on croplands. By providing continental level empirical analysis of a cropland area of 105 million hectares across 42 countries in Africa over a span of 15 years (starting from 2016), the fundamental objective was to align empirical data and economic valuation to help inform policy decisions in the future. The study provides detailed evidence that sustainable management of land and soil is economically a better alternative—as compared to degrade first and repair later. Thus, taking action against soil erosion from the 105 million hectares of croplands in 42 countries over the next 15 years will generate benefits of about 2.48 trillion USD or 62.4 billion USD per year in net present value.

With Agenda 2030, access to reliable data has become even more important than before. Monitoring of SDGs' targets like land degradation neutrality relies on data. Data is also needed to provide a sound knowledge base for the actual implementation of measures to reach the goals. Fortunately, data availability is not the main issue anymore. A wealth of data is available from the rapidly growing archives of earth observation satellites, as well as from research institutions and governmental and intergovernmental institutions. However, the pressing issue is the access to data and the sharing of data. At the international level, the European Commission's INSPIRE directive and the emerging Global Soil Information System of the Global Soil Partnership are two important initiatives to improve data access. At the national level, governments are facing the need to establish domestic spatial data infrastructures and to regulate the access to data.

Soil degradation is widely considered to be a key factor undermining agricultural livelihoods. Therefore, Germany welcomes and strongly supports the Land Degradation Neutrality (LDN) process of the UN Convention to Combat Desertification (UNCCD).

The development of concrete LDN targets in many countries leads to an increased awareness for the often-neglected issue of land and soil degradation. As of today, 113 countries have committed to set LDN targets and associated measures. Possible levers to reach LDN targets include legal instruments. Germany explicitly addresses LDN in its recent national sustainable development strategy.

Only close cooperation and partnership combined with a cross-sectoral perspective will lead to contextualized ways of reaching the LDN's ambitious vision. Sensible coordination of UNCCD activities with other agenda 2030 processes, the Rio Conventions, and other relevant initiatives is important, specifically to use synergies and to mitigate competitive situations among institutions.

Preservation of fertile soils worldwide is a prerequisite for the subsistence of mankind. Soil degradation leads to poverty, hunger, migration, and even conflicts. In Uganda, high population growth paired with massive deforestation and expansion of agricultural land and settlement zones contributes to urbanization, which in turn results in higher levels of unemployment, increased crime rates, and the formation of slums. Hence, land issues and their respective regulations are crucial.

Chargé d'Affaires a.i. of the Embassy   Petra Kochendörfer
of the Federal Republic of Germany
Kampala, Germany

Foreword

Setting the Agenda on Soil Governance in Africa and Uganda in the 21st Century

Soil is not only an essential resource for life but also at the heart of the main challenges faced by the planet including the production of food and energy and the delivery of essential environmental services.

As one of the land resources, soil is central to human well-being as it supports grazing, cultivation, energy provision, and settlement. Despite having significant relevance for food production, water, forests, and protection of biodiversity among others, soil is typically considered as a given resource rather than a separate resource in need of stewardship. Much thought is given to the different sectors, such as water, forests, and biodiversity, in total neglect of soil. This neglect has seen an increase in degradation of soils, which has caused negative repercussions to the prioritized sectors. Nevertheless, little has been done to change this situation.

It is imperative to note that soil has a fundamental role in mitigating the effects of climate change, in maintenance of water springs, and in sustaining biodiversity.

Communities and households continue to take land use decisions without due consideration of the delicate balance between productivity, ecosystem health, changing land uses, human welfare, poor inherent soil qualities, population pressure, insecure land tenure, and climate change among others. As a result, soil degradation and declining productivity is rampant.

The breakdown in land tenure and management systems has caused land degradation, which is evidenced by loss of vegetation, soil erosion, and soil fertility loss in most of the districts in Uganda. In spite of land tenure being a tool for soil conservation (since it involves sets of rules and regulations used to control and manage natural resources, biodiversity, and the general environment), landholding rights have been used to the detriment of the environment.

This is because the often human-focused rights and activities of landholders greatly affect the health and resilience of the soil. As a result, unsustainable land use practices have led to land degradation and have reduced the ecological and social resilience of landscapes. This is evident in the declining productivity, especially in connection with fragile ecosystems as well as the limited use of improved farming techniques and inputs to ensure soil conservation. Soil and land degradation stemming from deforestation, pollution from pesticides, burning of grasslands and organic residues, and continuous cultivation with minimum soil fertility enhancement leads to soil erosion and organic matter and nutrient depletion in these areas. Other unsustainable land-use practices, such as overgrazing and mechanical tilling, have produced compacted soil layers and often bare grounds in extreme cases in the country.

The decline of soil fertility in Africa is closely connected with steep increases in population. This has led to land fragmentation and a multiplicity of land ownership arrangements due to increased dwelling demands.

According to the International Center for Soil Fertility and Agricultural Development, there is substantial soil decline in every major region of sub-Saharan Africa, with the highest rates of depletion in Uganda, Guinea, Congo, Angola, Rwanda, and Burundi. Land degradation in Uganda ranges from 20% in relatively flat and vegetation-covered areas to 90% in the eastern and southwestern highlands.

Therefore, soil degradation is a major global problem. Its effects are felt most strongly in developing countries where large proportions of the population reap their livelihoods directly from the soil. Consequently, the maintenance of good soil quality is thus vital for the environmental and economic sustainability of land use activities—hence the need to effectively insert soil in public policies as a major sector like all the others to which it is connected.

The world and the African region have developed several international and regional instruments (which have been ratified by Uganda) that can be looked at in recognition of the importance of soil conservation. These build up a good debate for the inclusion of soil as a major sector in the environment.

The Convention on Biological Diversity (CBD) 1992 was ratified by Uganda in 1993. It aims at conserving biological diversity, promoting the sustainable use of its components, and encouraging equitable sharing of the benefits arising out of the utilization of genetic resources, soil being one of them.

The United Nations Framework Convention on Climate Change (UNFCCC) 1992 was signed by Uganda in 1992 and ratified in 1993. It was adopted to regulate levels of greenhouse gas concentration in the atmosphere to avoid the occurrence of climate change on a level that would compromise food production and to enable economic development to proceed in a sustainable manner. In 2002, Uganda also ratified the Kyoto Protocol 1998, a protocol to the UNFCCC.

The United Nations Convention to Combat Desertification (UNCCD) 1994 was signed by Uganda in 1996 and ratified in 1997. The convention accords the development of National Action Plans a central role in the strategy to combat desertification. The purpose of the National Action Plans is to identify the factors contributing to desertification and practical measures necessary to combat desertification. Article 8 of the Regional Implementation Annex for Africa urges states to include adjustment of institutional and regulatory framework of natural resource management to provide security of tenure for the local population in National Action Plans. Article 4 further sets out a number of general obligations for the signatory parties that include adopting an integrated approach addressing the physical, biological, and socioeconomic aspects of desertification. Uganda has about 30% drylands cover of its total land area stretching from the northeast to the southwestern part of the country (known as the cattle corridor). Such an instrument is relevant for soil protection in the case of Uganda's cattle corridor where desertification is a threat due to uncontrolled cattle grazing.

Domestically, Uganda has developed several policies and laws that are relevant to soil management. These include, among others, Uganda Vision 2040, the National Land Use Policy, the Land Policy, National Agriculture Policy 2013, Uganda Forestry Policy 2001, the Draft National Soils Policy for Uganda, the Prohibition of the Burning of Grass Act, Cattle Grazing Act, the National Environment Management Policy for Uganda (1994), the National Environment (Minimum Standards for Management of Soil Quality) Regulations, National Forestry and Tree Planting Act, Regulations on Mountainous and Hilly Areas, and Regulations on Wetlands, Riverbanks and Lakeshores. However, a lot more has to be done.

Fortunately, the international community is awakening to the threats concerning the soils of the planet and to the challenges that lie before us. Organizing a workshop like this is a good example. I have read through the program and noted that this workshop has an international dimension, which makes it important because many problems related to the soil are of a global nature. It will also allow us to learn from good experiences that have been adopted in other countries.

As we know, we are living a moment of construction of a post-2015 development agenda. I believe that all efforts should be made in order to include soil in the Sustainable Development Goals, thereby showing clearly the relevance and importance of good soil management.

Minister of Agriculture, Fisheries   Vincent Frerrio Bamulangaki Ssempijja
and Animal Industry, Entebbe
Republic of Uganda

Preface

Despite being the continent with the least amount of land degradation, the pressure on soils in Africa is currently enormous and continuously increasing due to a range of factors including poverty, overexploitation, population growth, and climate change.

The protection and sustainable management of soils is a precondition for sustainable development and even more so for the subsistence of humankind. Without ample areas of fertile soils, there is no food security and zero chance to mitigate climate change. Degraded soils result in hunger, famine, migration, and, under certain circumstances, even conflicts. This is particularly true for African states.

For example, attaining food security is particularly illusive because agricultural productivity remains low in Africa. Overstocking, overgrazing, water erosion, landslides, and over-application of agro-chemicals are impediments for sustainable agriculture. In addition, the poor section of the population heavily depends on land and other natural resources for immediate needs, which in turn is an additional driver for land degradation.

The Workshop on "Legal instruments for the effective protection and sustainable management of soils" was held at Hotel Africana, Kampala, from 26 to 27 September, 2017. It was jointly organized by the German Environment Agency, the "Climate Policy and Energy Security Program for Sub-Saharan Africa" of the Konrad-Adenauer Foundation, the Makerere University, and Kampala International University. The workshop was attended by about 50 experts from African countries and abroad as well as from the United Nations Convention to Combat Desertification (UNCCD), the Food and Agriculture Organization of the United Nations (FAO), the International Union for Conservation of Nation (IUCN), the United Nations Environmental Programme (UNEP), the African Soil Partnership, and the German Foundation for International Cooperation (GIZ). The workshop was a milestone for soil protection in Africa as it addressed the most critical regulatory issues and brought together key experts and actors on this topic.

This volume documents the discussions held at this workshop, publishing some of the presentations together with further substantially related chapters and the so-called Kampala outcome document, which was put together by the organizers of the workshop. The volume includes chapters on international and regional soil governance (Byron-Cox, Ginzky, and Kibugi), national soil regulation (Kasimbazi, Kassim), mining law (Sanni), and aspects concerning tenure rights (Mandi, Godard, and Orubebe), as well as on Australian soil law (Hannam). It culminates with the "Kampala outcome document," which captures the major outcomes of the workshop in Kampala.

We, the editors, would like to thank Liam Willoughby and Alyssa Bowra (University of New England), Mercy Teko (Strathmore University), and Johannes Ehlers and Pradeep Singh (Bremen University) for their timely, competent, and thorough language review of this book. In addition, special thanks go to Cornelia von Bohuszewicz, Germany, for the editing of the footnotes and the reference lists of all chapters. Finally, we extend our sincere appreciation to Ms. Anke Seyfried for her assistance in arranging this publication.

Kampala, Uganda   Hadijah Yahyah
Dessau, Germany   Harald Ginzky
Kampala, Uganda   Emmanuel Kasimbazi
Nairobi, Kenya   Robert Kibugi
Stellenbosch, South Africa   Oliver C. Ruppel
2019

Contents

From Desertification to Land Degradation Neutrality: The UNCCD and the Development of Legal Instruments for Protection of Soils

Richard Byron-Cox

1 Introduction

Land was and remains central to the state, which is the traditional, original and classic subject of Public International Law, that is to say, a state generally speaking must have some territory (land) of its own. Land is important to the state as it, inter alia, determines size of territory; its geo-political influence; and sometimes comes with resources ranging from gold and silver to the free labour of conquered peoples.

Today, land maintains its importance to the state for some of the aforementioned reasons, but there are additional very important factors including its social, environmental, and agricultural and agro-forestry values. Indeed, in this age of population explosion[1] and the consequences following in train including demand for more food and water; land's bio-productive value is now of paramount importance to human's and this planet's future existence. Hence, the need to protect and maintain this value.

In the context of the above, bio-productive lands essentially mean soils that are really the fundamental basis of living land. And humans' connection to and dependence on land cannot be overstated. As Joseph Conrad says, "Each blade of grass has its spot on earth whence it draws its life, its strength; and so is man rooted to the land from which he draws his faith together with his life."[2] Thus, there is a real and eternal value of living, bio-productive land to humankind. But have we, the "masters of the planet", understood this? Does our behavior towards the land on a global scale reflect an appreciation of its cardinal role in our everyday existence?

---

[1]See generally: Rosenberg (2019).
[2]Conrad (1900), p. 207.

R. Byron-Cox (✉)
Secretariat of the United Nations Convention to Combat Desertification, Action Programme Alignment and Capacity Building, Bonn, Germany
e-mail: RByronCox@unccd.int

© Springer Nature Switzerland AG 2020
H. Yahyah et al. (eds.), *Legal Instruments for Sustainable Soil Management in Africa*, International Yearbook of Soil Law and Policy,
https://doi.org/10.1007/978-3-030-36004-7_1

1

1.1   The Status of Soils Around the World

It is true that there are people who have long understood the value of bio-productive land, that is to say that good soils contribute to the wellbeing of humans—if not to the life of the planet as a whole.[3] However, soil and its productive capacity were not subjects that humanity as a whole concerned itself with so as to make it a priority for protection at the national and international levels. While this might sound like a sweeping statement, one just has to look at the latest "State of the World's Land and Water Resources" (SOLAW) report, which shows, inter alia, that 25% of the planet's lands is highly degraded, 8% moderately degraded and 36% stable or slightly degraded, while 18% is bare.[4] It is clear from this report that soil degradation is very serious indeed, that it has global effects, and that it must be arrested as a matter of necessary urgency. Indeed, the SOLAW report goes on to state that "large parts of all continents are experiencing degradation, with particularly high incidence of degradation down the west coast of the Americas, across Southern Europe and North Africa, across the Sahel and the Horn of Africa, and throughout Asia."[5]

This degradation did not happen overnight but is rather the consequence of a long process. And there is no disputing the role of human hands as a major cause agent in this degradation process.[6] Indeed, the available statistics[7] seem to suggest that we have had a terrible behaviour towards the earth. With us now needing more of everything (including water, food and energy), thereby creating greater and greater demand on the bio-productive lands of the planet, the question is: can we continue our present behaviour as regards our misuse and abuse of the land resources of this planet? If not, then we must change that behaviour at every level. More than that, the new and appropriate behaviour must be compulsory universally. This is why law is necessary for the protection of soils, for it is law that regulates the behaviour of humans in the modern state. So new laws are needed to prescribe a new and better relation of humans to land,[8] and this is absolutely for land is the very basis of all existence. Without a proper legal framework both at the international and national level that addresses this question of our behavioural change to one of protection of the planet's lands, the situation is bound to get worse.[9]

---

[3]Brevik (2005).

[4]FAO (2011), p. 111.

[5]FAO (2011), p. 111.

[6]Blaikie and Brookfield (1987).

[7]Eswaran et al. (2001).

[8]Boer and Hannam (2005).

[9]IPBES (2018).

2 The Universal Protection of Soils Comes into the Picture Via the SDGs

At its 70th session, the General Assembly of the United Nations adopted the 2030 Agenda[10] on Sustainable Development. In this document, many Heads of States of the world declared their intentions to pursue seventeen Sustainable Development Goals (SDGs), in order to ensure that the future of the planet is not compromised as we pursue present and future socio-economic development. These goals are to be achieved by the year 2030, and seek to address the fundamental question of balancing environmental protection with the needs of social and economic development. Along with these goals are their one hundred and sixty-nine targets which the international community will strive to achieve.

Goal 15 aims to "Protect, restore and promote sustainable use of terrestrial ecosystems, sustainably manage forests, combat desertification, and halt and reverse land degradation and halt biodiversity loss." The essence of this goal is securing terrestrial (land based) ecosystems services which can only be done through, inter alia, the protection of bio-productive land, or put another way, by the protection of soils.[11]

As with the other SDGs, this one also has its targets among which is Target 15.3 that declares, "By 2030, combat desertification, restore degraded land and soil, including land affected by desertification, drought and floods, and strive to achieve a land degradation-neutral world." So here we have a written declaration by the international community to address the question of soil protection by, inter alia, restoring "degraded land and soil." This was the first time the United Nations decided to take such action to protect the bio-productive land/soils of this planet as a whole and at such a level.

The question must, however, be asked of what is meant by a "land degradation-neutral world?" This is a pertinent question for it is logical that one must know what something is; i.e. one must first have a concept of it before one can try to achieve it. And very central to this first question is the question of from whence the concept of a land degradation-neutral world? Where did this idea come from? And why is the international community convinced that this idea should become a target of this all-important 2030 Sustainable Development Agenda on which the fate of this planet might very well be hinged? To answer these questions, it is necessary to visit the creation, history, evolution and practice of the implementation of the United Nations Convention to Combat Desertification (UNCCD).

---

[10]United Nations, General Assembly (2015) The 2030 Agenda for Sustainable Development.
[11]Adhikari and Hartemink (2016).

2.1 A Wake Up Call from the Sahel

The term desertification is, more often than not, completely misunderstood and is usually deemed to mean naturally occurring deserts. In fact, desertification has nothing to do with these deserts, all of which existed long before the first Homo sapiens evolved to set foot on this planet. The Namib Desert for example is 55 million years old and is a naturally occupying ecosystem. Consequently, when one speaks of desertification, one is speaking of bio-productive land becoming so unproductive due to several factors, including the improper actions and behaviour of humans towards the land, which causes it to take on desert-like features.[12] The UNCCD is designed to address this latter challenge of the protection and enhancement of bio-productive land. But how, and more importantly, why did the UNCCD come into being?

In the 1970s, serious and prolonged drought led to tragic consequences in the Sahel for humans, animals and the environment in general as the land failed to produce the food needed to sustain normal live and living. The international community could not simply sit back and watch the unfolding catastrophe of massive loss of human and animal life. Something had to be done. The response was the United Nations 1977 Conference on Desertification (UNCOD) held in Nairobi, Kenya, which culminated in the adoption of the Plan of Action to Combat Desertification (PACD).[13] It was soon realized that the PACD was not working too well; that desertification seriously affects other parts of the world; that the plan was too limited in scope and geopolitical reach, and that the consequences of desertification were global economic, social and environmental.

There was, therefore, a need for a global approach, response and commitment if a lasting solution was to be found. It must be noted here that even with these serious limitations, the Nairobi Conference was a signal moment when the international community, however slowly, began to realize that the death of soils could reap havoc in many parts of the world, leading to terrible catastrophes with dire global consequences. The UNCOD gave this question of desertification/land degradation some international currency as regards policy, which will eventually lead to the forming of law (UNCCD), and ultimately to the adoption of SDG Target 15.3.

By 1992, much of the peoples of the world became cognisant of the threat our present attitude to the pursuit of development poses to the environment and consequently to our very existence.[14] In keeping with this, they accepted that desertification, if not a global problem from the standpoint that not every nation suffers from it, was indeed a global problem due to its far reaching consequences for humanity as a whole. This gave birth to the concept of the need for a convention to address this issue. An intergovernmental group to work out the text of a global convention was

---

[12]Eswaran et al. (2001).

[13]Bauer and Stringer (2009).

[14]UNESCO (1992) The Rio Declaration on Environment and Development.

set up at Rio during the United Nations Conference on Environment and Development (UNCED) in 1992. The result of this group's work is the UNCCD.

The UNCCD is the first and only universal international legal instrument to date that addresses the question of the protection of land/soils in areas affected by desertification, land degradation and drought. It was this Convention that formally introduced understandings and concepts such as desertification and land degradation to international legal lexicon. And while the Convention as negotiated is full of blanket norms that offer little in terms of targets to be achieved and of indicators by which progress of implementation can or should be measured, it remains an international legal document focused on the protection of soils, providing a general international legal framework that offers broad guidance on this issue. And, it is the only one of its kind. By virtue of these two features, the UNCCD's objectives, principles, general norms and stipulations were and are great influences and fundamental stimuli in the generation of more international and national legislation and legal frameworks aimed at addressing the question of the protection of soils.[15]

This Convention has been in forced since 1996 and today has a total of 197 Parties, including the African and European Unions.

Its norms range from the international legal understanding of various terminologies important to and being use for the development of legal instruments for protection of soils, to actions—some within the legislative framework—that Parties are required to undertake in their efforts at implementation.[16]

This question of efficacious implementation became predominant once the Convention entered into force, and this is when and where the UNCCD (being the first and only international legal instrument of its kind) really found reflection in the activities of states as regards the protection of land including in the development of the relevant legislative frameworks needed for this protection to take place.

As a starting point in focusing their implementation efforts, those countries that have determined that they are affected by desertification or land degradation and want to actively address these issues developed National Action Programmes (NAPs).[17] These NAPs by nature are first and foremost policy documents and do not in many cases have the weight of legal authority. However, in some cases they are adopted by the parliaments, senates and other such legislative bodies, thus giving them the force of law. In other cases, they are established by the relevant state authority by way of decree or resolution. And still in others, convention or practice gives them the force of law, for once they emanate from a ministry or department of authoritative legal standing, they then have the power of decree. There are also cases

---

[15]Republic of Namibia (2004) Namibia's Third National Report on the implementation of the United Nations Convention to Combat Desertification.

[16]Articles 3, 4, 5 and 6 of the UNCCD.

[17]All NAPs are hosted on the UNCCD website at: https://www.unccd.int/convention/action-programmes.

where governments have passed special laws declaring the NAP or programmes related thereto as public policy that is to be followed.[18]

There are other ways where the UNCCD has had if not a direct and powerful influence of legislative frameworks that directly protect soils, then it would have at least had and is having indirect effects. These include legislation regarding women rights to land,[19] addressing the issues of sand and dust storms, and the development of regional and sub-regional plans to address the questions of desertification and land degradation.[20]

At the global level, the UNCCD as a process helping to drive legislation to support the protection of soils did not stop with the negotiation of the text of the Convention.

On the contrary, Parties have adopted various decisions since its adoption—which apart from being legal norms at the international level, find reception in the national laws of many countries, especially via their NAPs and other policies and laws designed to protect soils. Additionally, these decisions help to broaden the scope of the entire process of soil protection by covering many related areas, including issues of science (such as soil science) and policy (such as land tenure) and the science-policy interface.[21] The last of these three is extremely relevant in this regard because for policy to be effective it must be informed by science (facts), and for it to have true authority, it needs the backing of law.

Notwithstanding all supra dictum, there have been serious challenges with the implementation of the UNCCD.[22] One reason for this was that the issue of desertification was still not seen in many quarters as a global problem. However, with the constant increase in land degradation taking place in many parts of the globe, it is clear that sustainable land management is not an issue purely for the arid and semiarid lands but for all lands; it has serious consequences not just for life on the planet across the board but also for the life of the planet itself as we know it. A different (broader) approach of addressing land degradation globally was therefore needed, and the UNCCD elected to initiate this process, which culminated in what is known today as Land Degradation Neutrality (LDN).

Most activities for social and economic development require the use of land, which in many cases leads to the serious depletion of the land's productive capacity or contributes to land degradation. This reduction in the size, and the damage and even destruction of bio-productive lands through the change of use for other purposes (such as mining, industrial development and housing) can also be caused -and is indeed caused in many instances by other activities—which at a prima facie glance may not seem to be the case. Activities like some practices of crop farming and animal

---

[18]China Committee for the Implementation of the UNCCD (2006).

[19]Mor (2018).

[20]See Regional Annexes of the UNCCD.

[21]United Nations (2013) United Nations Convention to Combat Desertification. Report of the Conference of the Parties on its eleventh session, held in Windhoek from 16 to 27 September 2013.

[22]Ambalam (2014).

husbandry are cases in point. The way these are sometimes done has led to negative effects, including various forms of the depletion soil productivity and even to ultimate soil exhaustion.[23]

These facts are some of the fundamental reasons that justified the UNCCD paying attention to the question of land degradation as a global phenomenon, rather than just something that takes place in the arid and semiarid lands.

With a growing world population and the consequent demand on land to produce more of everything it offers, the logical question is: Can we continue to damage and destroy our bio-productive land on such a scale?

The answer to this is certainly no, that is if we care about sustaining life on this planet, and indeed the life of the planet itself. This then leads to a second logical question: So what do we do? It is in trying to answer this second question that the UNCCD came up with the concept of Zero Net Land Degradation.[24]

3   The Concept of Zero Net Land Degradation

The elementary, yet very powerful, idea behind Zero Net Land Degradation is that social and economic development can—and should—proceed apace without any further loss in the bio-productive lands of the planet. This is now universally accepted, as reflected in goal 15 of the Sustainable Development Goals. Put simply, it means that using any year as the base point of measurement (whatever year that may be), all efforts would be made to ensure that as a global measure, no more land is degraded through our social or economic activity or from any other action of humans. And if this loss indeed must unavoidably take place, then an equal amount of degraded land must be restored to its bio-productivity capacity. This, therefore, was the genesis of a process where addressing land degradation would ultimately become a central element in efforts to achieve sustainable land management and ultimately sustainable development.

Zero Net Land Degradation proved to be a very hard concept to sell for a multitude of reasons, not least among which was the question of how to implement and measure it.[25] There were other serious questions of legal and political nature, such as whether the UNCCD was going beyond its mandate in seeking to introduce this concept the UNCCD. In short, the concept of Zero Net Land Degradation did not receive universal or majority acceptance in the international arena.

---

[23]European Communities (2009).

[24]United Nations Convention to Combat Desertification, UNCCD (2012) Zero Net Land Degradation - A Sustainable Development Goal for Rio+20.

[25]Stavi and Lal (2014).

However, new reports on land degradation are painting a graver and graver picture of the unfolding situation. Moreover, the Aichi Targets[26] and instruments like the Paris Agreement established in very vivid terms the direct connection between bio-diversity protection and land, and climate change and land respectively.[27]

The nations of the world knew that the game was up. Tackling land degradation is not an option, but rather a necessity if we are to achieve global sustainable development and efficacious global environmental governance.

Thus, dealing with land degradation is crucial to meeting the objectives of sustainable development and the 2030 Agenda. The question was therefore: if not Zero Net Land Degradation then what?

3.1   Rio + 20 to the Rescue

Thankfully, there were those who realized that while Zero Net Land Degradation might not be acceptable to all and might not be the perfect answer, it was still an idea from which something very constructive could be built as regards defining some of the things sustainable land management should be about.[28] Still, this was not enough. It had to become part of the international environment and, more importantly, sustainable development agenda proper if states were actually going to view it as something of real importance. The gateway for this was provided by the adoption of the Rio + 20 outcome document entitled "The Future We Want," during the United Nations Conference on Sustainable Development (UNCSD) held in Rio de Janeiro in 2012.

Under the subheading "Desertification, land degradation and drought," paragraphs 205 to 209 of this document focuses on states addressing the challenges of land degradation as these relate to sustainable development. So, in paragraph 205, the international community clearly states, inter alia, that "We recognize the economic and social significance of good land management, including soil, particularly its contribution to economic growth, biodiversity, sustainable agriculture and food security, eradicating poverty, women's empowerment, addressing climate change and improving water availability. We stress that desertification; land degradation and drought are challenges of a global dimension and continue to pose serious challenges to the sustainable development of all countries [. . .]".

All these are central themes of the UNCCD and underline its direct and powerful influence on the development of this part of the document. As will be shown later

---

[26]See UNEP, Convention on Biological Diversity (2010) Strategic Plan for Biodiversity 2011–2020 and the Aichi Targets (COP 10 Decision X/2) specifically Targets 4, 5, 7, 11 and 14.

[27]United Nations, UNFCC (2015) Paris Agreement. See especially Article 5.2.

[28]United Nations (2011) Convention to Combat Desertification. Report of the Conference of the Parties on its tenth session, held in Changwon from 10 to 21 October 2011.

herein, this redounded to influence significantly the development and acceptance of Goal 15 as part of Agenda 2030.

Moreover, in paragraph 206 the document continues to add that, "We recognize the need for urgent action to reverse land degradation. In view of this, we will strive to achieve a land-degradation neutral world in the context of sustainable development."

Here, we find for the first time in an internationally adopted policy document lexicon referring to land degradation neutrality (LDN), and a commitment by states (even if non-binding), to make efforts to achieve it.

Turning to paragraph 207, the Parties state that, "We reaffirm our resolve in accordance with the United Nations Convention to Combat Desertification to take coordinated action nationally, regionally and internationally, to monitor, globally, land degradation [...]. We note the importance of mitigating the effects of desertification, land degradation and drought, including by [....] restoring degraded lands, improving soil quality [...]".[29]

3.2 Pushing the Envelope of LDN

The UNCCD realized that these paragraphs in "The Future We Want" presented a unique opportunity to place the problematic of addressing the issue of land degradation neutrality in the front and center of the international sustainable development agenda. Thus, it set about trying to do just that.

In little over a year after the adoption of the Rio + 20 outcome document with its paragraphs on addressing the issue of land degradation, and more specifically, the intention to "strive to achieve a land-degradation neutral world," the UNCCD had prepared and was ready to implement a pilot project on LDN.

This project's main aim was to show that achieving LDN in practice is indeed possible.[30] Crucially, this project looked at important issues such as setting LDN baselines, possible indicators for measuring LDN, setting targets to achieve LDN, and integrating LDN into national priorities and commitments.

Fourteen countries from around the world participated in this project. Their final reports all indicate that achieving LDN, while challenging, was not an impossibility.[31] It must be noted that this project played a key role in helping to establish the three main indicators for measuring progress with achieving LDN,[32] which is first and foremost about protecting soils.

---

[29]United Nations, General Assembly (2012) A/RES/66/288- The Future we want.

[30]UNCCD (2016a) Scaling up Land Degradation Neutrality Target Setting - from Lessons to Actions: 14 Pilot Countries' Experiences.

[31]UNCCD (2016c) Land Degradation Neutrality (LDN) Pilot Project country reports.

[32]UNCCD (2016b) Achieving Land Degradation Neutrality at the Country Level: Building Blocks for LDN Target Setting.

It is clear from this that UNCCD, as a global process fashioned upon an international agreement (namely the Convention itself), is truly the foundation on which these focused efforts to protect soils (including in the legislative realm) are being built both at the international and national levels. The international community could not but acknowledge this fact when adopting the SDGs by including in the title of Goal 15, "combat desertification, and halt and reverse land degradation," which is practically a quote on the very essence of the UNCCD.

3.3 The Evolution Continues: The Significance of Agenda 2030

As already noted, "The Future We Want" is a non-binding document. However, its significance in driving the sustainable development process and all that it entails, including the LDN process at all levels especially in the global policy spheres, is evident. Indeed, it forms part of the fundamental basis for the adoption of SDG 15, and in relation to the specific subject of this essay, the adoption of Target 15.3. Agreement on this target was a key piece of the puzzle as regards making LDN a universal objective, which would be reflected in the behaviour of states both internationally and nationally including in their legislation.

Since the adoption of SDG 15 and its Target 15.3, there has been and continues to be great interest in its implementation.

While this implementation is possible without a legal framework, it is far better if such a framework is established at all levels, that is to say international and national.

The framers of Agenda 2030 clearly understood that the adoption of Agenda 2030 must provide the international policy basis and general consensus on which specific legal norms and frameworks can be and must be built in specific areas such the protection of soils, if the goals are to be achieved. This postulation is confirmed by taking a look at the sub-section entitled "Means of Implementation" and in particular paragraph 45. Here, the intention to give legal sanction to these goals and targets is made pellucid with the Parties declaring that, "We acknowledge also the essential role of national parliaments through their enactment of legislation [. . .] for the effective implementation of our commitments." Herein then lies the importance of the adoption of the SDGs as an international policy document, which plays a foundation role in the further development of legal frameworks to support sustainable development.

The agreement on Target 15.3 has had and continues to have a tremendous positive effect on the developments in both science and policy/law as pertains to the protection of soils.[33] Following the adoption of this target, Parties have determined that the UNCCD shall be the "custodian" for helping all countries in their effort to achieve LDN.

---

[33]See e.g. Decision 18/COP.14 of the final report of the UNCCD COP 14.

As part of efforts to realise this mandate, the UNCCD has set up the Global Mechanism an LDN Target Setting Programme (TSP) through its operational arm. This programme offers countries wishing to work on implementing LDN various forms of assistance. What is most important to note here, however, is that countries actively undertake measures including legislative measures to promote sustainable land management and thereby the protection of soils through this programme.[34]

Policy is generally more effective when its implementation is backed by the authority of law and when there is a proper legislative framework in which it can operate. It is still early days yet for many countries as regards the development of national legislation aimed specifically at the protection of soils. But many countries have begun the process, being guided by the requirements for achieving the sustainable development goals and the international framework being laid out for the same. The UNCCD as an international legal instrument and as a global process continues to be in the vanguard of soil protection. It has and continues to push the envelope as regards the protection of soil in various ways. It is only a matter of time, as the history of this process to date has shown, before more legislation at both the international and national levels shall emerge to ensure better protection of soils in the future.

4   Conclusion

This chapter demonstrates that while the objective of having proper legislative frameworks to protect soils universally is still a long way off, the UNCCD as a legal instrument and a living process has and continues to be a stimulus and guide for this purpose. Of course, the UNCCD is only part of this process and needs to continue to work with and within other processes, both at the national and international levels, if the objective is to be achieved. As regards these other processes, actions in the spheres of climate change, biodiversity and the continuing development of international environmental law among others, are some of the areas vital to the further evolution of legislative frameworks to enhance soil protection. The UNCCD has been working synergistically with these and other relevant processes at all levels, and with it being the "custodian" for LDN, this cooperation will not only continue, but will certainly intensify, as it fulfils its role as vanguard in efforts to ensure universal soil protect.

References

Adhikari K, Hartemink AE (2016) Linking soils to ecosystem services — a global review. Geoderma 262:101–111

---

[34]United Nations (2017).

Ambalam K (2014) Challenges of compliance with multilateral environmental agreements: the case of the United Nations Convention to Combat Desertification in Africa. J Sustain Dev Stud 5 (2):145–168

Bauer S, Stringer LC (2009) The role of science in the global governance of desertification. J Environ Dev 18(3):248–267

Blaikie PM, Brookfield H (eds) (1987) Land degradation and society. Routledge

Boer B, Hannam I (2005) Developing a global soil regime. Int J Rural Law Policy (IJRLP) 2015 1:4123

Brevik EC (2005) A brief history of soil science. Land use, land cover and soil sciences, vol VI

China Committee for the Implementation of the UNCCD (2006) China National Report on the Implementation of the United Nation's Convention to Combat Desertification. Retrieved from: http://www.unccd-prais.com/Uploads/GetReportPdf/cbc38bc4-5cbe-4312-aace-a0fa014a4aa7

Conrad J (1900) Lord Jim. Blackwood's Magazine

Eswaran H, Lal R, Reich PF (2001) Land degradation: an overview. Responses to Land Degradation. Proceedings of the 2nd International Conference on Land Degradation and Desertification, Khon Kaen. Oxford Press, New Delhi

European Communities (2009) Sustainable agriculture and soil conservation. Fact Sheet No. 1: linking soil degradation processes, soil-friendly farming practices and soil-relevant policy measures. Retrieved from: https://esdac.jrc.ec.europa.eu/projects/SOCO/FactSheets/EN%20Fact%20Sheet.pdf

Food and Agriculture Organization of the United Nations, FAO (2011) The state of the world's land and water resources for food and agriculture (SOLAW)- Managing systems at risk. Food and Agriculture Organization of the United Nations, Rome and Earthscan, London, p 111. Accessed 24 Sept 2019

Intergovernmental Science-Policy Platform on Biodiversity and Ecosystem Services, IPBES (2018) Media Release: Worsening Worldwide Land Degradation Now 'Critical', Undermining Well-Being of 3.2 Billion People. Retrieved from: https://www.ipbes.net/news/media-release-worsening-worldwide-land-degradation-now-%E2%80%98critical%E2%80%99-undermining-well-being-32

Mor T (2018) Towards a gender-responsive implementation of the United Nations Convention to Combat Desertification. UN Women Headquarters. Retrieved from: https://www.unwomen.org/en/digital-library/publications/2018/2/towards-a-gender-responsive-implementation-of-the-un-convention-to-combat-desertification

Republic of Namibia (2004) Namibia's Third National Report on the implementation of the United Nations Convention to Combat Desertification. Retrieved from: http://www.unccd-prais.com/Uploads/GetReportPdf/5de56af3-5e16-4828-b1bd-a0fa014a4a80

Rosenberg M (2019) Current world population and future projections. ThoughtCo, 22 Aug 2019. http://www.thoughtco.com/current-world-population-1435270

Stavi I, Lal R (2014) Achieving zero net land degradation: challenges and opportunities. J Arid Environ 112:44–51

United Nations (2011) Convention to Combat Desertification. Report of the Conference of the Parties on its tenth session, held in Changwon from 10 to 21 October 2011. Part one: Proceedings. Retrieved from: https://www.unccd.int/official-documents/cop-10-changwon-2011/iccdcop1031

United Nations (2013) United Nations Convention to Combat Desertification. Report of the Conference of the Parties on its eleventh session, held in Windhoek from 16 to 27 September 2013. Part one: proceedings. Retrieved from: https://www.unccd.int/official-documents/cop-11-windhoek-2013

United Nations (2017) Convention to Combat Desertification. Report of the Conference of the Parties on its thirteenth session, held in Ordos, China from 6 to 16 September 2017. Part one: Proceedings, pp 13, 21–22. Retrieved from: https://www.unccd.int/official-documents/cop-13-ordos-china-2017/cop1321

United Nations Convention to Combat Desertification, UNCCD (2012) Zero Net Land Degradation - A Sustainable Development Goal for Rio+20. Retrieved from: https://sustainabledevelopment.un.org/index.php?page=view&type=400&nr=526&menu=35

United Nations Convention to Combat Desertification, UNCCD (2016a) Scaling up Land Degradation Neutrality Target Setting - from Lessons to Actions: 14 Pilot Countries' Experiences. UNCCD publications. Retrieved from: https://www.unccd.int/publications/scaling-land-degradation-neutrality-target-setting-lessons-actions-14-pilot-countries

United Nations Convention to Combat Desertification, UNCCD (2016b) Achieving Land Degradation Neutrality at the Country Level: Building Blocks for LDN Target Setting, UNCCD Publications. Retrieved from: https://www.unccd.int/publications/achieving-land-degradation-neutrality-country-level-building-blocks-ldn-target-setting

United Nations Convention to Combat Desertification, UNCCD (2016c) Land Degradation Neutrality (LDN) Pilot Project country reports. Retrieved from: https://www.unccd.int/publications/scaling-land-degradation-neutrality-target-setting-lessons-actions-14-pilot-countries

United Nations Educational, Scientific and Cultural Organization, UNESCO (1992) The Rio Declaration on Environment and Development

United Nations Environmental Program (UNEP), Convention on Biological Diversity (2010) Decision adopted by the Conference of the Parties to the Convention on Biological Diversity in its tenth meeting, Nagoya, Japan, 18–29 October 2010 (COP 10 Decision X/2. Retrieved from: https://www.cbd.int/decision/cop/?id=12268

United Nations, General Assembly (2012) A/RES/66/288- The Future we want

United Nations, General Assembly (2015) Transforming our World: The 2030 Agenda for Sustainable Development

United Nations, United Nations Framework Convention on Climate Change, UNFCCC (2015) Paris Agreement

In Salvation of African Soils: Exploring the Window of Opportunity Through Nationally Determined Contributions to Implement Sustainable Soil Management

Robert Kibugi

1 Introduction

According to the World Soil Charter, "soils are fundamental to life on earth but human pressures on soil resources are reaching critical limits."[1] Soils have also been identified as the "foundation of food production and many essential ecosystem services."[2] Resolution 3/6 of the Third Session of the United Nations Environment Assembly (UNEA-3) affirmed this noting that soil is one of the largest reservoirs of biodiversity, but the negative impacts of the contamination of soil are undermining productivity and sustainability of ecosystems, biodiversity, agriculture and food security, and clean ground and surface water."[3]

Although soil is deemed the foundation of food production and many ecosystem services, the 2019 Global Resources Outlook[4] reported that "food production is responsible for the majority of biodiversity loss, soil erosion and a large share of anthropogenic greenhouse gas emissions."[5] In addition, soils are facing degradation, and pollution from various sources. The Second Global Chemicals Outlook, released in 2019, reports that "soils throughout the world are contaminated by hazardous chemicals, heavy metals and certain pesticides."[6] This is affirmed by the Sixth

---

[1]FAO (2015) Revised World Soil Charter, Preamble.

[2]FAO (2017) Voluntary Guidelines for Sustainable Soil Management.

[3]UNEA (2017) Managing soil pollution to achieve sustainable development - UNEP/EA.3/Res.6, p. 1.

[4]UNEP, International Resource Panel (2019) Global Resources Outlook 2019.

[5]UNEP, International Resource Panel (2019) Global Resources Outlook 2019, p. 17.

[6]UNEA, Global Chemicals Outlook II: summary for policymakers - UNEP/EA.4/21, p. 8.

R. Kibugi (✉)
School of Law, University of Nairobi, Nairobi, Kenya

© Springer Nature Switzerland AG 2020
H. Yahyah et al. (eds.), *Legal Instruments for Sustainable Soil Management in Africa*, International Yearbook of Soil Law and Policy,
https://doi.org/10.1007/978-3-030-36004-7_2

Global Environmental Outlook (Geo-6) Regional Assessment for Africa, which reports that blanket fertilizer application without assessing the soil needs produces negative results,[7] with countries such as Ethiopia embarking on soil fertility mapping to determine the appropriate type of fertilizer for a particular soil type in order to mitigate against the inappropriate use of fertilizer.[8] The 2030 Sustainable Development Agenda affirms the importance of proper management of soils. This is evident, for instance, in SDG 3.9 which aims, by 2030, to substantially reduce the number of deaths and illnesses from soil pollution and contamination. SDG 12.4 also aims to substantially reduce the release of chemicals and their wastes into soil in order to minimize their adverse impacts on human health and the environment.

In this context, UNEA-3 Resolution 3/6 observed that contamination of soil is hampering the achievement of the Sustainable Development Goals, including Goals including Goals 1, 2, 3, 6, 12, 13 and 15."[9] In a nutshell, these SDGs concern the following global goals:

a) SDG 1 aims to end poverty in all its forms everywhere.
b) SDG 2—aims to end hunger, achieve food security and improved nutrition and promote sustainable agriculture
c) SDG 3—aims to ensure healthy lives and promote well-being for all at all ages
d) SDG 6—aims to ensure availability and sustainable management of water and sanitation for all
e) SDG 12—aims to ensure sustainable consumption and production patterns
f) SDG 13—calls on countries to take urgent action to combat climate change and its impacts
g) SDG 15—calls for action to protect, restore and promote sustainable use of terrestrial ecosystems, sustainably manage forests, combat desertification, and halt and reverse land degradation and halt biodiversity loss

The sustainable management of soil is therefore critical, as from the foregoing discussions, the health of soils impacts up to seven of the SDGs directly. The attainment of sustainable soil management is the theme this chapter, and it is identified by the revised World Soil Charter, as one if its principles, as follows:

> Sustainable management of global soil resources is critical to meeting increased societal demands in a responsible manner. Soil management is sustainable if the supporting, provisioning, regulating, and cultural services provided by soil are maintained or enhanced without significantly impairing the soil functions that enable those services.[10]

Based on the foregoing, sustainable soil management is concerned with ensuring there is a balance between the supporting and provisioning services for plant

---

[7]UNEP (2016) GEO-6 Regional Assessment for Africa, p. 39.

[8]UNEP (2016) GEO-6 Regional Assessment for Africa, p. 39.

[9]UNEA (2017) Managing soil pollution to achieve sustainable development - UNEP/EA.3/Res.6, p. 1.

[10]FAO (2015) Revised World Soil Charter, Preamble.

production and the regulating services the soil provides for water quality and availability and for atmospheric greenhouse gas composition is a particular concern."[11]

This chapter examines the nexus between soil governance and climate change impacts, looking at the window of opportunity presented by Nationally Determined Contributions (NDCs) under the 2015 Paris Agreement, for African countries to consolidate and enhance sustainable soil management, as part of their climate strategies and commitments. Section 2 of the chapter explores the link between soil governance and climate change impacts. Section 3 reviews the African approach to soil governance, focusing on treaties, to show there is a continent-wide concern on soil health. Section 4 ventures into the opportunity window provided by NDCs reviewing the legal nature of these NDCs. Specifically, this section argues that under a new form of Common But Differentiated Responsibilities (CBDR), African nations have submitted NDCs based on their respective capabilities and national circumstances. This means that African nations, in designing their NDCs, have the leeway to identify and prioritize sustainable soil management actions as part of their NDC mitigation and adaptation commitments. Once these are submitted, they become binding legal commitments. African nations can thus utilize the financing and technology transfer windows available under the Paris Agreement, for the benefit of soil management as part of climate action. Section 5 reviews a random selection of ten NDCs submitted to the Secretariat of the United Nations Framework Convention on Climate Change in 2015, reviewing how these NDCs address soil management. Section 6 is the conclusion of the chapter.

2 The Nexus Between Soil Governance and Climate Change Impacts

According to the Voluntary Guidelines for Sustainable Soil Management[12] soil is the world's largest terrestrial pool of carbon,[13] about twice that found as carbon dioxide in the atmosphere.[14] Soils therefore play a critical role in regulating climate and mitigating climate change through trade-offs between greenhouse gas emission and carbon sequestration.[15] Historically, soils in managed ecosystems such as farmlands have lost a portion of this carbon through land use change, some of which has remained in the atmosphere.[16]

---

[11]FAO (2015) Revised World Soil Charter, Preamble.

[12]FAO (2017) Voluntary Guidelines for Sustainable Soil Management.

[13]FAO (2017) Voluntary Guidelines for Sustainable Soil Management, p. 1.

[14]Smith (2012).

[15]FAO (2017) Voluntary Guidelines for Sustainable Soil Management, p. 8.

[16]Smith (2012).

Soil organic matter plays a central role in maintaining soil functions and preventing soil degradation and is strategic for climate change adaptation and mitigation.[17]

Consequently, a loss of soil organic carbon due to inappropriate land use or the use of poor soil management or cropping practices can cause a decline in soil quality and soil structure, and increase soil erosion, potentially leading to emissions of carbon into the atmosphere.[18] On the other hand, appropriate land use and soil management can lead to increased soil organic carbon and improved soil quality that can partially mitigate the rise of atmospheric CO_2. There are limitations of soil as a carbon sink, based on soil type, or selected land use (e.g. intensive mechanized land tilling, or slash and burn) with a risk of leakage thus focus on permanence would be needed. This is important because sustainable soil management focuses on the complete set of ecosystem services provided by soils. Where optimally under-taken, sustainable soil management contributes towards achievement of SDG 13, to combat climate change, and SDG 15.3 to combat desertification, restore degraded land and soil, including land affected by desertification, drought and floods, and strive to achieve a land degradation-neutral world.

According to the Sixth Global Environmental Outlook Report (GEO-6) Regional Assessment for Africa, the continent has 60% of the world's unconverted arable land, but land productivity remains low in the region as a result of mineral poor soils and land degradation caused by inappropriate farming practices, deforestation, mining activities, and desertification.[19] In terms of farming, Geo-6 noted that cultivation in much of Africa encroaches on environmentally fragile areas such as steep slopes, riverbanks, shallow soils and wetlands, often without appropriate conservation measures in place, leading to increased soil erosion.[20] Small holder farmers use their land continuously with no rotation, resulting in declining crop yields and the loosening and washing away of soil exposed to natural forces such as wind and water.[21] Indeed, Agriculture production in Africa is hampered by the predominance of inherently low soil fertility, and fragile ecosystem, which is exacerbated by the fact that over 60% of the soil types in Africa represent hot, arid or immature assemblages.[22]

There is also a high prevalence of dry lands that are more susceptible to land degradation and desertification, in addition to extensive coverage of land with low soil resilience, which refers to soil that is permanently damaged from degradation.[23]

This suggests that African soil, and land are inherently vulnerable to experience more severe consequences of climate change. The Geo-6 report for African endorses

---

[17]FAO (2017) Voluntary Guidelines for Sustainable Soil Management, p. 8.

[18]FAO (2017) Voluntary Guidelines for Sustainable Soil Management, p. 8.

[19]UNEP (2016) GEO-6 Regional Assessment for Africa, p. 37.

[20]UNEP (2016) GEO-6 Regional Assessment for Africa, p. 47.

[21]UNEP (2016) GEO-6 Regional Assessment for Africa, p. 47.

[22]Ginzky et al. (2017), p. 390.

[23]Ginzky et al. (2017), p. 390.

this view, noting that climate will have serious implications on the availability of arable land and freshwater. In addition, Africa's vulnerability to the impacts of climate change is worsened by its comparatively low adaptive capacity.[24] Consequently, climate change will have direct impacts on food provisioning services on the continent,[25] which is a problem for soil governance because the maintenance of the provisioning services provided by soil is one of the elements of sustainable soil management. As an ecosystem asset, soil provides an additional ecosystem service through carbon capture, such as through mangrove ecosystems.[26]

Mangrove forests, occurring between the high and low water marks along many coastlines, as well as seagrasses, are important to carbon capture and climate regulation The role of mangroves (and other aquatic ecosystems) as carbon sinks is commonly referred to as blue carbon.[27] Although they occupy less than 2% of the world's ocean surface area, blue carbon ecosystems are estimated to bury nearly 10% of the yearly estimated organic carbon burial in the oceans.[28] In real terms, mangroves have a mean whole-ecosystem carbon stock of 956 tons of carbon per hectare compared with 241 tons of carbon per hectare, for rain forests.[29]

Unlike many terrestrial systems that store organic primarily in living biomass, vegetated coastal ecosystems store much of their organic carbon stocks in the soil sediments, which may produce carbon sinks of hundreds to thousands of years age.[30]

Regardless of forest size therefore, soils constitute the largest carbon pool, with the percentage of the total soil pool varying from 44% for rain forests to 70% for peat swamps, 75% for mangroves, and over 90% for marshes and seagrasses.[31] However, this stored organic carbon risks being released back to the atmosphere when blue carbon ecosystems are degraded.[32]

Even with the importance of soils in climate regulation clear, the continuing pollution of soil from economic activities and other sources (such as chemical fertilizers and pesticides) will lower resilience of African soils increases the vulnerability of these soils to climate change impact. Yet production processes continue to generate significant chemical releases, waste and hazardous waste to the soil.[33] UNEA-3 Resolution 3/6 acknowledged this noting that while soils are known to be an essential element for climate change mitigation and resilience, soil pollution leads to a reduction in soil biological activity therefore reducing the capacity of soil

---

[24]UNEP (2016) GEO-6 Regional Assessment for Africa, p. 111.

[25]UNEP (2016) GEO-6 Regional Assessment for Africa, p. 112.

[26]UNEP (2016) GEO-6 Regional Assessment for Africa, p. 105.

[27]IUCN (2016) National Blue Carbon Policy Assessment Framework.

[28]Githaiga et al. (2017).

[29]Alongi (2014).

[30]Githaiga et al. (2017).

[31]Alongi (2014), p. 199.

[32]Githaiga et al. (2017), p. 1.

[33]UNEA (2019) Global Chemicals Outlook II: summary for policymakers - UNEP/EA.4/21, p. 8.

to act as a carbon sink.[34] Taking into account the inherent vulnerability of most soils in Africa, reliance of smallholder farming, low adaptive capacity, and generally higher vulnerability of Africa to the impacts of climate change, there is need for soils to be managed sustainably. This means ensuring that economic activities do not harm soil capability to provide the provisioning, regulating, supporting, cultural ecosystem services. In addition, the outcomes of sustainable soil management in Africa should enhance soil carbon capture capacity, on land and ocean environments. This helps regulate GHG emissions, and enhances resilience of soils, which can then contribute to fulfilment food security, and the various SDG commitments. The world currently has an enhanced focus on climate change actions, and therefore integrating sustainable soil management into climate change priorities of African nations utilizes a valuable window of opportunity.

3   Assessment of the African Approach to Soil Governance

At the continental level in Africa, there are various legal instruments whose provisions govern, or impact governance of soil. The 1968 African Convention on the Conservation Nature and Natural Resources[35] (Algiers Convention) contains explicit provisions on soil governance.

It requires States to take effective measures for conservation and improvement of soil, with a particular focus on combating erosion and misuse of the soil.[36] The Algiers Convention requires African States to do this by establishing land-use plans based on scientific investigations (ecological, pedological, economic, and sociological) and, in particular, classification of land-use capability.[37] The Algiers Convention is cognizant of the role of agriculture in Africa, and requires African States to take certain actions when implementing agricultural practices and agrarian reforms, as follows:

1. Improve soil conservation and introduce improved farming methods, which ensure long-term productivity of the land, and;
2. Control erosion caused by various forms of land-use which may lead to loss of vegetation cover.

It is important to highlight that the fundamental principle of the Algiers Convention is adoption of the necessary to ensure conservation, utilization and development of soil, water, flora and faunal resources in accordance with scientific principles, and,

---

[34]UNEA (2017) Managing soil pollution to achieve sustainable development - UNEP/EA.3/Res.6, p. 1.
[35]1968 African Convention on Nature and Natural Resources.
[36]1968 African Convention, Article 4.
[37]1968 African Convention, Article 4(a).

with due regard to the best interests of the people of Africa.[38] The Algiers Convention failed to provide the institutional structures that would have facilitated its effective implementation. It also did not establish mechanisms to encourage compliance and enforcement.[39]

In 2003, African nations signed a Revised African Convention on the Conservation of Nature and Natural Resource (Revised African Convention).[40] The objectives of the Revised Convention are (i) to enhance environmental protection; (ii) foster the conservation and sustainable use of natural resources; and (iii) to harmonize and coordinate policies in these fields.[41] The third objective is important, especially because soil governance in Africa requires effective harmonization of sectoral policies across countries, and within countries, to minimize and eliminate the deleterious effects of sectoral actions. The Revised Convention defines natural resources to include soils.[42]

Specifically, the Convention provides for measures to protect land and soil, requiring Parties to take effective measures to prevent land degradation, and adopt measures for the conservation and improvement of the soil in order to combat erosion, soil misuse and the deterioration of the soils physical, chemical and biological or economic properties.[43] The Convention further requires African nations, when implementing agricultural practices and agrarian reform, to take the following measures[44]:

(i) improve soil conservation and introduce sustainable farming and forestry practices, which ensure long-term productivity of the land,
(ii) control erosion caused by land misuse and mismanagement which may lead to long-term loss of surface soils and vegetation cover,
(iii) control pollution caused by agricultural activities, including aquaculture and animal husbandry.

Further, Parties are required to prevent soil erosion, pollution or any other form degradation, when implementing non-agricultural public works including mining and waste disposal. Where land degradation has occurred, the Convention requires African nations to plan and implement mitigation and rehabilitation measures.[45]

Although the Revised African Convention does not make provision for climate change actions, it requires implementation of measures that would lessen the vulnerability of African soils, and sustain the soils ability to provide ecosystem services. In Sects. 1 and 2 above, the chapter highlighted the challenges facing soil health in

---

[38] 1968 African Convention, Article 2.
[39] IUCN (2004).
[40] African Union (2017) Revised African Convention.
[41] Revised African Convention, Article 2.
[42] Revised African Convention, Article 5.
[43] Revised African Convention, Article 6.
[44] Revised African Convention, Article 6.
[45] Revised African Convention, Article 6.

Africa, and how low soil resilience and low adaptive capacity increases the vulnerability of the soil to adverse impacts of climate change. There is need to enhance application of sustainable soil management in order to restore and sustain the ability of soil to provide provisioning (e.g. food production), supporting (e.g. biodiversity), regulating (e.g. carbon capture, climate control), and cultural ecosystem services. The Revised African Convention provides a wide scope of legal authority and mechanism to address soil erosion, degradation, pollution, and damage to the physical, chemical, or biological properties of soil. Its provisions resonate with the Sustainable Development Goals, including SDGs 2, 6, 13, and 15.

Although African nations approved the Revised African Convention in 2003, the Algiers Convention remains applicable for all nations that have not ratified to the 2003 Convention. This is based on article 34 of the Revised African Convention which provides, with respect to its relationship with Algiers Convention, that "between Parties which are bound by this (Revised African) Convention, only this Convention shall apply." The Revised Convention came into operation on 23 July 2016 upon attaining the 15th ratification by Burkina Faso. At the time of writing, the Convention has 17 Ratifications, which is very low considering 44 out of 54 African countries have appended signature. Widespread implementation of the valuable provisions of the Convention will therefore remain low at 17 out of 54 African countries. In addition, the Revision Convention does not have a direct correlation with climate change, although climate regulation is a major function of soil, in terms of sustainable soil management.

Based on the foregoing assessment, it is evident that continental level treaty law recognizes the important place of sustainable soil governance, but that both the Algiers and Maputo Conventions face implementation challenges. Thus this pathway will, by itself, not provide the needed window of opportunity for African nations to prioritize soil governance towards sustainable outcomes where soils provide balanced ecosystem services. In the next section, the chapter therefore reviews how the global climate change framework through the Paris Agreement could be utilized as a window through which African nations could prioritize action on soils, by integrating this into their national climate change priorities. The focus is on the role that NDCs can play.

4 Legal Nature of Nationally Determined Contributions (NDCs) Under Paris Agreement and Utility to Soil Governance

It is important point out that the Paris Agreement[46] in its provisions for NDCs has modified the principle of common but differentiated responsibilities (CBDR). Through article 4 of the Paris Agreement, the CBDR principle have evolved, in

---

[46]United Nations, UNFCCC (2015) Paris Agreement.

practice, from the point where developing countries bore no obligations under the Kyoto Protocol (absence of universal participation), to a system under the Paris Agreement that permits voluntary definition of national targets (considering respective capabilities, national circumstances, and national objectives) when then translate into binding legal commitments.[47]

In this case, all countries participate in defining their commitments with no legal differentiation between developed and developed countries, except in terms of how countries define their NDC commitments.[48] As will be seen below, African countries have a focus in defining how their mitigation actions catalyse or relate to adaptation benefits, because adaptation actions address socio-economic and environmentally systemic challenges and therefore respond to poverty reduction and developmental priorities.

Therefore the Paris Agreement requires each Party to prepare and communicate NDCs that it intends to achieve.[49] On the basis of the submitted NDC, each Parties is required to pursue domestic mitigation measures in order to achieve the defined national objectives.[50] According to the Paris Agreement, developed country Parties should continue to take the lead by undertaking economy-wide absolute emission reduction targets.[51] Developing country Parties on the other hand should continue enhancing their mitigation efforts while moving, over time, towards economy-wide emission reduction or limitation targets.[52] It is important to highlight that under the Paris Agreement, Parties are required to communicate the NDCs in every five year period,[53] and each Party's successive NDC should represent a progression beyond that Party's [then] current NDC contribution.[54] This progression should reflect the highest national ambition, reflecting the country's common but differentiated responsibilities and respective capabilities, in light of different national circumstances.[55]

This represents a system where developing African nations are actively participating in implementation of the Paris Agreement by defining their nationally determined contributions toward mitigation of GHGs through mitigation. Developing country Parties will do so by defining their own national objectives, in light of their specific national circumstances and capabilities. On this basis, many developing countries will set out mitigation mechanisms for reduction of emissions, but as part of their NDCs they also set out adaptation measures they consider crucial to their national objectives and respective capabilities.

---

[47]Kibugi (2018).

[48]Ferreira (2017).

[49]Paris Agreement, 2015, Article 4(2).

[50]Paris Agreement, 2015, Article 4(2).

[51]Paris Agreement, 2015, Article 4(4).

[52]Paris Agreement, 2015, Article 4(4).

[53]Paris Agreement, 2015, Article 4(9).

[54]Paris Agreement, 2015, Article 4(3).

[55]Paris Agreement, 2015, Article 4(3).

For this reason, when examining how countries have treated the governance of soil in their NDC submissions, it is important to pay attention to both their mitigation and adaptation submissions.

5 Assessing Prioritization and Integration of Soil Governance in Nationally Determined Contributions Submitted by African Countries

In this section, the chapter presents a succinct review of randomly selected ten NDCs submitted by African nations to the UNFCCC. In the review, the chapter focuses on how the NDCs have treated soil governance, relative to climate change actions, in context of both mitigation and adaptation. Where provided, explanations are also set out below. The NDCs have been obtained from the Interim NDC Register operated by the Secretariat of the UNFCCC.[56]

5.1 Prioritization and Integration of Soil Governance by Algeria in Its NDC

In its NDC,[57] Algeria defines its own national objectives and capabilities and sets out national circumstances. The national objective is to achieve reduction of greenhouse gas emissions by 7% to 22%, by 2030, compared to a business as usual (BAU) scenario, conditional on external support in terms of finance, technology development and transfer, and capacity building. The 7% GHG reduction will be achieved with national means.[58]

The NDC stipulates that through its mitigation actions for by 2030, considering its socio-economic development objectives, and taking into account its national circumstances, Algeria will contribute, on an equitable basis, to the achievement of the objective of article 2 of the UNFCCC.[59]

In its national circumstances, Algeria details that it is affected by desertification and land degradation and most of the country is arid or semi- arid.[60] More than 50 million of hectares face highly deteriorated conditions and due to this land degradation, rural farmers and breeders have been forced into exodus to large cities,

---

[56]The Interim NDC Register is available online: https://www4.unfccc.int/sites/ndcstaging/Pages/Home.aspx.

[57]Algeria NDC, 2015.

[58]Algeria NDC, 2015, p. 6.

[59]Algeria NDC, 2015, p. 8.

[60]Algeria NDC, 2015, p. 4.

for survival.[61] Algeria identifies actions on land and soil as part of adaptation measures, noting that priority will be given to the protection of the population and the preservation of natural resources against risk of extreme events.[62]

For this reason, Algeria's adaptation objectives include actions to "fight against erosion and rehabilitate degraded lands as part of the efforts to combat desertification."[63] The country further notes that it will mainstream action on climate change impacts into sectoral strategies, in particular for agriculture, water management, public health and transport.[64]

5.2   Prioritization and Integration of Soil Governance by Botswana in Its NDC

In its NDC, Botswana's national objective is to achieve an overall emissions reduction of 15% by 2030, taking 2010 as the base year.[65] The NDC specifies that as a semi-arid country, Botswana is vulnerable to the impacts of climate change and places high priority on adaptation to reducing vulnerability.[66] The NDC identifies extreme droughts based on low rainfall and soil conditions as most common in south-western Botswana.[67] In its NDC, Botswana announced it was developing a National Adaptation Plan that prioritizes Climate Smart Agriculture that includes techniques such as zero-tillage, multi-cropping to increase mulching which in turn reduces soil erosion.[68]

5.3   Prioritization and Integration of Soil Governance by Nigeria in Its NDC

In its NDC,[69] Nigeria identifies soil erosion as a major risk factor that lowers adaptive capacity. The NDC notes that climate change related heavier than normal rainfall is expected to worsen soil erosion, which is already at catastrophic levels in the southern part of the country.[70] It points out recent increase in landslides as

---

[61]Algeria NDC, 2015, p. 4.

[62]Algeria NDC, 2015, p. 8.

[63]Algeria NDC, 2015, p. 8.

[64]Algeria NDC, 2015, p. 8.

[65]Botswana NDC, 2015, p. 1.

[66]Botswana NDC, 2015, p. 2.

[67]Botswana NDC, 2015, p. 4.

[68]Botswana NDC, 2015, p. 2.

[69]Nigeria, NDC, 2015.

[70]Nigeria, NDC, 2015, p. 5.

evidence of possible climate change-induced changes in erosion intensity.[71] The NDC further prioritizes the adoption of better soil management practices as a key strategy for agricultural crops and livestock sectors.[72] Further, the NDC makes reference to the Nigerian National Agricultural Resilience Framework (NARF 2014), that articulates policy options, opportunities and required interventions to among others, improve farming practices and productivity through land and water management strategies such as soil fertility enhancement, and soil erosion control.[73]

5.4   Prioritization and Integration of Soil Governance by Egypt in Its NDC

The NDC for Egypt[74] focuses both on adaptation and mitigation actions. In terms of mitigation, concerning GHG emissions reduction in non-energy sector, the NDC prioritizes agriculture, among others. For agriculture, the mitigation measures are merely listed down as (i) manure management, (ii) agricultural soils, and (iii) field burning of agricultural residues.[75] It is worth to emphasize that the action concerning "agricultural soils" as a mitigation measure is merely listed down, without further details on the specific actions to be taken. Presumably this could include soil management practices to reduce GHG emissions (e.g. conservation agriculture techniques) or enhance the carbon sequestration capacity of the agricultural soils. In terms of adaptation, the NDC focuses relevant actions on the agricultural sector proposing to increase efficiency of irrigation water use while maintaining crop productivity and protection of land from degradation.[76] In addition, Egypt commits to review land use policies and agricultural expansion programs to take into account possibilities of land degradation in Nile Delta and other affected areas resulting from Mediterranean Sea level rise.[77]

---

[71]Nigeria, NDC, 2015, p. 5.
[72]Nigeria, NDC, 2015, p. 7.
[73]Nigeria, NDC, 2015, p. 8.
[74]Egypt, NDC, 2015.
[75]Egypt, NDC, 2015, p. 7.
[76]Egypt, NDC, 2015, p. 7.
[77]Egypt, NDC, 2015, p. 8.

5.5 Prioritization and Integration of Soil Governance by Zambia in Its NDC

Zambia's NDC[78] proposes mitigation and adaptation actions that specifically prioritize soil management. In terms of mitigation, the NDC focuses on sustainable agriculture options, such as conservation or smart agriculture that also result in adaptation benefits.[79] The expected outcomes of this approach include reduced GHG emissions due to reduced fertilizer use and less turning of soil, enhanced soil carbon sequestration and improved soil productivity.[80] The same focus on sustainable agriculture is prioritized as an adaptation action, where climate smart agriculture practices are prioritized. They include conservation agriculture, agroforestry, use of drought tolerant varieties, water use efficiency management and fertilizer use efficiency management.[81] These adaptation actions are expected to result in (i) increased soil fertility and conservation leading to improved crop productivity; as well as (ii) improved agro-biodiversity conservation.[82]

5.6 Prioritization and Integration of Soil Governance by Rwanda in Its NDC

Rwanda's NDC[83] identifies priority adaptation actions that have mitigation co-benefits, and reports that Rwanda is vulnerable to climate change as evidenced by experiences of seasonal shortages in food supply as a result of poor harvests caused by droughts and flooding and soil erosion.[84] According to the NDC, in order to adapt to this situation, Rwanda is mainstreaming agro-ecology technologies in its current agriculture intensification programme, to ensure total adoption of sustainable food production by 2030.[85] The NDC lists the mitigation co-benefit for this action as the reduction of GHG emissions from land use change.[86] The agro-ecological techniques include spatial plant stacking, agro-forestry, kitchen gardens, nutrient recycling, and water conservation.[87]

---

[78]Zambia, NDC, 2015.

[79]Zambia, NDC, 2015, p. 3.

[80]Zambia, NDC, 2015, p. 3.

[81]Zambia, NDC, 2015, p. 7.

[82]Zambia, NDC, 2015, p. 7.

[83]Rwanda, NDC, 2015.

[84]Rwanda, NDC, 2015, p. 3.

[85]Rwanda, NDC, 2015, p. 3.

[86]Rwanda, NDC, 2015, p. 3.

[87]Rwanda, NDC, 2015, p. 3.

The NDC also proposed utilizing resource recovery and reuse through implementing country-wide organic waste composting for all food production, in order to restore and soil fertility.[88] Additionally, the NDC proposed enrichment of compost with inorganic fertilizers, in order to minimize the intensive use of inorganic fertilizers as they have adverse environmental impacts.[89] According the NDC, the effectiveness of composted organic waste can be further improved by enriching and blending it with nutrients (such as Nitrogen phosphorus), which ensures a more efficient use of inorganic fertilizers, and adds valuable organic matter to soils, which also maximizes terrestrial carbon in farm soils.[90] The mitigation co-benefit for this action is reduction of GHG emissions from fertilizer manufacturing processes.[91]

Rwanda also identified mainstreaming sustainable past management techniques as a soil-relevant action by implementing the push-pull system of using napier grass and desmodium legume to manage pests under maize, sorghum, millets and rain- fed rice plantations.[92] The main adaptation benefits of the push-pull system are the increase of yields, soil fertility improvement through nitrogen fixation and provision of a continuous supply of fodder.[93]

The NDC identified reduced GHG emissions from enteric fermentation as a mitigation co-benefit of this action. The Rwanda NDC also identified soil conservation and land husbandry is a key adaptation action.[94] It reported that 90% of Rwanda's crop land is on slopes ranging from 5 to 50% which makes it vulnerable to climate change impacts like soil erosion leading to permanent fertility loss.[95] As a result, the NDC indicated that Rwanda would expand its soil conservation and land husbandry programmes through installation of land protection structures like radical and progressive terraces, and intensive agro-forestry targeting all arable land by 2030.[96]

5.7 Prioritization and Integration of Soil Governance by Ethiopia in Its NDC

In its NDC,[97] Ethiopia identifies actions concerning soil management for adaptation, as long- and medium-term actions to address drought. This includes taking action to enhance ecosystem health through ecological farming, sustainable land management

---

[88]Rwanda, NDC, 2015, p. 4.

[89]Rwanda, NDC, 2015, p. 4.

[90]Rwanda, NDC, 2015, p. 4.

[91]Rwanda, NDC, 2015, p. 4.

[92]Rwanda, NDC, 2015, p. 5.

[93]Rwanda, NDC, 2015, p. 5.

[94]Rwanda, NDC, 2015, p. 5.

[95]Rwanda, NDC, 2015, p. 5.

[96]Rwanda, NDC, 2015, pp. 5–6.

[97]Ethiopia, NDC, 2015.

practices and improved livestock production practices to reverse soil erosion, restore water balance, and increase vegetation cover, including drought tolerant vegetation.[98] The NDC also commits Ethiopia to create biodiversity movement corridors, especially up towards higher terrain, in areas where most of the land is under cultivation.[99] The biodiversity corridors are intended to minimize biodiversity loss through enabling the re-establishment and movement of plant and animal species and varieties to areas suitable for their survival when temperature rises.[100]

5.8 Prioritization and Integration of Soil Governance by Malawi in Its NDC

The Malawi NDC[101] identifies mitigation options for agriculture that involve soil management options. These include improved manure management; promoting agroforestry systems in targeted locations as source of biomass and soil carbon sequestration.[102] Other actions include reducing use of synthetic fertilizers, encouraging application of organic amendments to fertilizers such as manure and crop residues that contain the potential to contribute to soil carbon levels.[103]

The NDC also prioritizes the planting of nitrogen fixing plants to reduce fertilizer usage; as well as potentially reduced and zero tillage.[104] These mitigation measures suggested in the agricultural sector will unconditionally contribute 100 Gg CO_2 equivalent mainly from reduced synthetic fertilizer application, and around 400 Gg CO_2 equivalent per annum from implementing climate smart agriculture extensively by 2040, conditional upon support.[105] However, it is important to note that according to the NDC, Malawi considers the overall mitigation potential of the agriculture sector to be comparably small.[106] Nonetheless, Malawi's NDC does not set out soil management as an adaptation action, or set out the adaptation co-benefits of the agriculture mitigation actions.

---

[98]Ethiopia, NDC, 2015, p. 6.

[99]Ethiopia, NDC, 2015, p. 6.

[100]Ethiopia, NDC, 2015, p. 6.

[101]Malawi, NDC, 2015.

[102]Malawi, NDC, 2015, p. 5.

[103]Malawi, NDC, 2015, p. 5.

[104]Malawi, NDC, 2015, p. 5.

[105]Malawi, NDC, 2015, p. 6.

[106]Malawi, NDC, 2015, p. 6.

5.9 Prioritization and Integration of Soil Governance by Tanzania in Its NDC

In its NDC, Tanzania reports that it is a net carbon sink, since out of a total 88 million hectares of land areas, 48.1 million are forested land and under different management regimes, current estimated total of 9.03^2 Trillion tons of carbon stock.[107] While there is no direct reference to soil governance, the NDC sets out adaptation actions in agriculture that could impact soil management. These include up-scaling the level of improvement of agricultural land and water management; and implementation of climate smart agriculture to increase crop yields.[108]

5.10 Prioritization and Integration of Soil Governance by Kenya in Its NDC

Kenya's NDC[109] indicates that more than 80% of the country's landmass is arid and semi-arid land (ASAL), and that the country has a low GHG emissions profile of 73 MtCO2eq in 2010, out of which 75% are from the land use, land-use change and forestry (LULUCF) and agriculture sectors.[110] Although the NDC makes no direct reference to soil management, Kenya commits to mainstream climate change adaptation in land reforms; to enhance the resilience of ecosystems to climate variability and change; and to promote climate smart agriculture.[111]

The Country has put in place as Climate Smart Agriculture Strategy[112] which identifies decline in soil health and water quality due to climate change and extreme weather events as a threat to climate smart agriculture. The strategy is one of the avenues for NDC implementation, and prioritizes actions for sustainable natural resources management to enhance resilience and adaptive capacity through among other actions, implementation of integrated soil health management that includes soil nutrient management, soil and water conservation, and conservation agriculture; restoration of degraded soils and conservation of soil biodiversity.[113]

---

[107]Tanzania, NDC, 2015. p. 3.

[108]Tanzania, NDC, 2015. p. 4.

[109]Kenya, NDC, 2015.

[110]Kenya, NDC, 2015, p. 1.

[111]Kenya, NDC, 2015, p. 5.

[112]Government of the Republic of Kenya (2018a) Kenya Climate Smart Agriculture Strategy 2017–2026.

[113]Government of the Republic of Kenya (2018a) Kenya Climate Smart Agriculture Strategy 2017–2026, p. 58.

5.11 Intermittent Conclusion

The foregoing analysis demonstrates that African countries have an inbuilt consciousness on the proximate relationship between land management, and soil management with their climate change action. Some of the NDCs reviewed above demonstrated a focus on soil, such as the case of Rwanda. Other NDCs focused only on land management, but listed out actions that will directly impact soil, such as biodiversity corridors. In several instances, countries listed out soil management actions as mitigation measures, while other countries listed them as adaptation measures. A number of instances directly correlated the co-benefits of either adaptation or mitigation actions that touched on agriculture or soil management. Regardless of the approach, this demonstrates that African countries are conscious of the link between climate change and soil management and have put forward their planned actions through NDCs.

6 Conclusion and Recommendations

Although African countries could be implementing soil and sustainable land management measures related to other conventions, such as the United Nations Convention to Combat Desertification, or the Convention on Biological Diversity, there is no cross-reference of this in the NDCs. More explicitly, the NDCs do not cross- refer to the actions African countries are taking through National Biodiversity Strategies and Action Plans (NBSAPs), which many countries are preparing a second version, at the time when they are unbundling implementation of NDCs.

Additionally, the NDCs do not cross-refer to treaty obligations of African countries either under the 1968 African Convention on Conservation of Nature and Natural Resources, or the 2003 revised version. As discussed earlier, both of these African Conventions very specifically provide for the management of soil. The concept of sustainable soil management, and its goal of maintaining soil functionality to provide ecological services that include carbon capture is not addressed directly in the NDC, although elements of it are evident. The 2003 Revised African Convention is now in force, but with 17 ratifications out of 54 African Countries. Nonetheless, implementation should commence so that Africa-specific soil actions can be integrated, with a link to commitments by African countries under the Paris Agreement, through the NDCs.

The chapter reviewed the NDCs as submitted the UNFCCC in 2015, as the only versions publicly available. It is possible that many countries have taken further steps nationally to implement specific soil management steps for enhancement of carbon stocks, with the co-benefits of improved soil health and productivity. Kenya, for instance, has taken additional action through its Climate Smart Agriculture

Strategy, recognizing that land degradation includes the loss of the various ecosystem services provided by soils.[114] This recognition is in harmony with the sustainable soil management approach proposed by the Revised World Soils Charter discussed earlier. The Climate Smart Agriculture Strategy further reports that due to pressure from agriculture, both soil and water resources are being degraded through erosion, salinization, seawater intrusion; and groundwater depletion, and that the current model of intensive agriculture is associated with a high carbon and greenhouse gas footprint, and highly vulnerable to the predicted impacts of climate change.[115] Taking this further, Kenya implements its NDC through a National Climate Change Action Plan (NCCAP) that is mandated by the 2016 Climate Change Act[116] as the tool for mainstreaming climate change actions across all sectors of Kenya's economy. The draft 2018–2023 NCCAP prioritizes increasing food and nutrition security, and indicates that within that period, Kenya will implement climate smart agricultural practices by increasing land under sustainable land management for agricultural production.[117]

It is estimated that this action will support the reclamation of 60,000 ha of degraded land; increase land area under integrated soil nutrient management increased by 250,000 acres while implementing minimum or no tillage conservation agriculture over the same acreage; and increase agro-forestry coverage by 200,000 acres.[118]

In addition to building resilience and enhancing adaptive capacity of food, nutrition and income security, these actions are expected to enhance mitigation through GHG emission reductions of 0.55 MtCO2e by 2022 (conservation tillage), and 1.66 MtCO2e by 2022 (agroforestry).[119] 2018 capacity building forum for African nations in context of the Comprehensive African Agriculture Development Programme and implementation of NDCs on agriculture reiterated that agriculture has a key role to play in climate change adaptation and mitigation, especially in developing countries.[120] There is therefore opportunity for African countries to prioritize actions on adaptation and mitigation with soil management, to deliver adaptation and mitigation co-benefits. In so doing, African nations should focus also on transboundary actions concerning neighbouring and contiguous countries to ensure that wrong action or non-action by neighbours does not have deleterious effects. This will also consolidate implementation of the 2030 Sustainable Development Agenda.

---

[114]Government of the Republic of Kenya (2018a) Kenya Climate Smart Agriculture Strategy 2017–2026, p. 50.

[115]Government of the Republic of Kenya (2018a) Kenya Climate Smart Agriculture Strategy 2017–2026, p. 50.

[116]Government of the Republic of Kenya (2016) Climate Change Act, No. 11 of 2016, section 13.

[117]Government of the Republic of Kenya (2018b) Draft National Climate Change Action Plan 2018–2023, p. 45.

[118]Government of the Republic of Kenya (2018b) Draft National Climate Change Action Plan 2018–2023, p. 45.

[119]Government of the Republic of Kenya (2018b) Draft National Climate Change Action Plan 2018–2023, p. 45.

[120]CGIAR (2018).

References

African Union (1968) African Convention on Nature and Natural Resources

African Union (2017) Revised African Convention on the Conservation of Nature and Natural Resources

Alongi DM (2014) Carbon cycling and storage in mangrove forests. Annu Rev Mar Sci 6:195–121

Consultative Group on International Agricultural Research, CGIAR (2018) Training workshop report: implementing Nationally Determined Contributions (NDC) Commitments in Agriculture. Retrieved from: https://ccafs.cgiar.org/publications/training-workshop-report-implementing-nationally-determined-contributions-ndc#.XYoVVy4zZaR

Ferreira PG (2017) From justice to participation: the Paris Agreement's pragmatic approach to differentiation. In: Abate RS (ed) Climate justice: case studies in global and regional governance challenges. Environmental Law Institute, Washington DC

Ginzky H et al (eds) (2017) International Yearbook of Soil Law and Policy 2017. Springer International Publishing, pp 387–411

Githaiga MN, Kairo JG, Gilpin L, Huxham M (2017) Carbon storage in the seagrass meadows of Gazi Bay, Kenya. PLoS ONE 12(5):e0177001

Government of the Republic of Kenya (2016) Climate Change Act, Kenya Gazette Supplement No. 68 (Acts No. 11)

Government of the Republic of Kenya (2018a) Kenya Climate Smart Agriculture Strategy 2017–2026

Government of the Republic of Kenya (2018b) National Climate Change Action Plan (Kenya): 2018–2022. Ministry of Environment and Forestry, Nairobi

International Union for Conservation of Nature and Natural Resources, IUCN (2004) An Introduction to the African Convention on the Conservation of Nature and Natural Resources, IUCN Environmental Policy and Law Paper No. 56

International Union for Conservation of Nature and Natural Resources, IUCN (2016) National Blue Carbon Policy Assessment Framework. Towards effective management of coastal carbon ecosystems. Retrieved from: https://portals.iucn.org/library/sites/library/files/documents/2016-080.pdf

Kibugi R (2018) Common but differentiated responsibilities in a North-South context: assessment of the evolving practice under climate change treaties. In: Faure M (ed) Elgar encyclopedia of environmental law, Chap. VI.44, pp 613–626

Smith P (2012) Soils and climate change. Curr Opin Environ Sustain 4(5):539–544

United Nations Environment Assembly of the United Nations Environment Programme, UNEA (2017) Managing soil pollution to achieve sustainable development- UNEP/EA.3/Res.6

United Nations Environment Assembly of the United Nations Environment Programme, UNEA (2019) Global Chemicals Outlook II: summary for policymakers. Report of the Executive Director. Fourth session, Nairobi, 11–15 March 2019. Available at: https://undocs.org/UNEP/EA.4/21

United Nations Environment Programme (2016) GEO-6 Regional Assessment for Africa. http://wedocs.unep.org/handle/20.500.11822/7595

United Nations Environmental Programme (UNEP), International Resource Panel (2019) Global Resources Outlook 2019: Natural Resources for the Future We Want

United Nations Food and Agriculture Organization, FAO (2015) Revised World Soil Charter

United Nations Food and Agriculture Organization, FAO (2017) Voluntary Guidelines for Sustainable Soil Management

United Nations, United Nations Framework Convention on Climate change (2015) Paris Agreement

Good Governance for "Sustainable Management of Soil" on National and International Level: How to Do It?

Harald Ginzky

1 Introduction

Soils are an issue which increasingly attracts the awareness of policy makers, scientists and the civil society. Three facts underline this assumption: the world's community has obliged itself in the 2030 Agenda for Sustainable Development of 2015[1] to strive to achieve, by 2030, a "land degradation neutral world".[2] Between the years 2013 to 2017, four "Global soil weeks" took place in Berlin, Germany.[3] UNCCD has interpreted itself as the international lead organization to deal with the objective of "land degradation neutrality".[4] The food crisis, the envisaged need for soils for renewable energy production about 10 years ago and the then initiated marketing of soils itself which was debated under the term "land grabbing", sparked the increasingly intensified debate on soils and lands.[5]

---

[1]United Nations (2015).

[2]As soils are the fundamental component of land the obligation for "land degradation neutrality" is valid for soils too.

[3]Further information under https://globalsoilweek.org/.

[4]Boer et al. (2016), pp. 61–62.

[5]Ginzky (2016), pp. 1–2. Also other environmental problems have been or are even still neglected: A problem which is now high on the agenda is marine litter in the oceans. But it took same time to get there. Currently, the loss of insects (in Germany long-term studies document a loss of about 75%) is actually not really debated although it might have tremendous effects on the food chain, the biodiversity and on food security.

H. Ginzky (✉)
German Environment Agency, Dessau, Germany

Development and Rule of Law Programme [DROP], University of Stellenbosch, Stellenbosch, South Africa
e-mail: harald.ginzky@uba.de

© Springer Nature Switzerland AG 2020
H. Yahyah et al. (eds.), *Legal Instruments for Sustainable Soil Management in Africa*, International Yearbook of Soil Law and Policy,
https://doi.org/10.1007/978-3-030-36004-7_3

Soils have been "neglected" over a long period of time.[6] Thus it could be stated that in most legislation the protection and sustainable management of water and air issues have been dealt with in specific laws, long before a comparable regulation was put in place for soils.[7] There are many explanations: almost every human activity degrades land/soil, soils are not considered as common good, as they are usually privately owned, soils/land are a core element of state sovereignty, the ecological services of soils have only been recognized just recently.

The chapter briefly shows that the sustainable management of soils is an urgent and demanding challenge of global nature and that the world's community has agreed to consider soils as such a global challenge. To deal with this challenge, good soil governance has to be in place. Therefore, the chapter goes on to explain that good soil governance should be designed with regard to the drivers and the specific threats of soil degradation. In this context, by way of drawing on German soil legislation, it is shown that German regulatory concept follows this driver/threat approach. Furthermore, existing pieces of legislation in Germany are analysed whether they foresee an effective implementation of the objective of "land degradation neutrality".

The chapter turns to the question how future international soil governance should be designed—given that there is a political will and momentum for new international soil provisions. To this end, the chapter explains that current international law does not provide an efficient governance for sustainable management of soils. It is argued, that it is recommendable to arrange for a cooperation between states and for a better coordination between existing treaties. Additionally, the chapter argues that in order to achieve measurable benefits for soils, it would be recommendable to draft the obligations as specific as politically possible with regard to drivers and threats. The chapter then shows that a new treaty consisting of a framework convention and subsidiary annexes which should deal with specific drivers would be recommendable. The chapter concludes with a short summary of the results and an outlook.

2 Sustainable Management of Soils as a Global Challenge

Soils have received attention and interest of the international community only in the first years of this century. Before that, the conservation of soils has not been regarded as a priority. Soils have been perceived as the poor cousin of water, air and biodiversity.[8]

The food crisis at the beginning of this century—and all its consequences—has put soils on the international agenda for the first time. It became clear that fertile soils

---

[6]Heuser (2017).

[7]For example: In Germany the first Soil Protection Act got in force 1998. Water protection regulation have been in place since the 1950th.

[8]Heuser (2017).

are a precondition for a sustainable development. Without fertile soils, security of food production is at risk. Moreover policy makers saw that fertile soils are needed for the production of plants for renewable energy, too. Scientific evidence showed that fertile soils are the second largest biological sequester of carbon and soils host an almost infinite biodiversity.[9]

Soils provide the following services which are indispensable for sustainable development:

- Basis for food production and production of plants for renewable energy
- Sequester for carbon and therefore important in the fight of climate change
- Host for biodiversity
- Vital role in world's biological cycle by *inter alia* storage for nutrients or filter for groundwater from hazardous substances
- Cultural and biological archive

Looking at the current state of soils worldwide, figures of the ongoing or even acceleration of land degradation are alarming. About one third of world's soils are already affected.[10] An area of the size Italy has been and will be—if nothing happens—constantly degraded per year.[11]

Soils are generally immobile and local. They are mostly privately owned.[12] However they provide services which are transboundary in nature as these services are essential for sustainable development.[13] Thus from a factual point of view sustainable management of soils could be regarded as global challenge.[14]

Anja Eikerman has branded the very illustrative term "hybrid good" for pointing out this duplicity with regard to forests.[15] Forests and soils are very comparable. Both forests and soils are local and generally immobile. Both are usually privately owned. At the same time both forests and soils provide essential services for sustainable development. Thus term "hybrid good" seems to be also an appropriate "concept" for soils.

Scholars have raised the question whether sustainable management of soils or land degradation is a "global concern of humankind", a legal principle which at least

---

[9]Ginzky (2016), p. 2.

[10]Ginzky (2016), p. 3.

[11]Linz and Lobos (2016), p. 197.

[12]"Privately owned" encompasses collective rights on land as also in this case there is a community holding rights on a specific piece of land.

[13]Land/soil degradation could cause hunger, poverty, migration and political or even military conflicts. Ginzky (2016), p. 7; Boer et al. (2016), p. 68; see also Flasbarth (2016), p. 17 mentioning that even the Syrian conflict has been at least partly caused by land degradation processes.

[14]The discussion has been so far focusing whether land degradation is a common concern of mankind. However, as soils are the fundamental component of land and degradation is the contrary to sustainable management, this previous debate is also relevant for soils or the conservation of fertile soils.

[15]Eikermann (2014), p. 4; Eikermann (2017), p. 417.

requires mandatorily cooperative actions by states.[16] In legal terms, so far, neither sustainable management of soils nor land degradation has been established as a "global concern of humankind".[17]

However, the international community has agreed to consider the sustainable management of soil and land as a global challenge. This "agreement" could be regarded as a political, but not legal commitment. Two documents are most important in this regard. First, the Outcome Document of Rio+20 which was jointly adopted by the UN General Assembly, states that "desertification, land degradation and drought are challenges of a global dimension and continue to pose serious challenges to the sustainable development of all countries".[18] Second, Target 15.3 of the Sustainable Development Goals stipulates that states should strive to achieve a "'land degradation neutral world' by 2030.[19]

3   Good Regulation for Soil Protection: Driver and Threat Specific

Soils are a global challenge both factually as well as politically. To deal with this challenge requires good governance.

The next section provides a conceptual approach how in general terms a good regulation to ensure a sustainable management of soils should be designed. The particular perspective is the national level. It first attempts to point out the particularities which have to be considered with regard to soil governance. Second, the conditions for a good soil governance will be explained. In a short excursus the general structure of soil regulation in Germany is presented and compared with these conditions of good soil governance. It should be highlighted that fragmentation per se is not a hint for ineffective soil governance.

3.1   Particularities of Soil Governance

In the following, the particularities of soil governance will be explained. They need to be taken into account when designing good soil governance.

---

[16]Ginzky (2017) and Boer (2014).

[17]Ginzky (2017), p. 433.

[18]United Nations (2012) The Future We Want, Para 2005.

[19]United Nations (2015).

Soils are at first a practically ubiquitous environmental media. Almost any human activity necessarily uses land and soil.[20] And almost any use of soil will cause degradation—at least to some extent.

Soil governance is a cross-cutting issue because it needs to be considered with regard to all human activities. There are certain main drivers of degradation of soils[21]: Agriculture, urbanization, land take for infrastructure such as streets or railways, industrial uses.

In addition soils are vulnerable to different types of threats. The EU Thematic Strategy for Soils of 2006[22] has differentiated eight types, including: soil erosion, decline in organic matter, local and diffuse contamination, sealing, compaction, and decline in biodiversity, salinization, floods and landslides.[23] Each of these threats requires a different management regime.

Some drivers of soil degradation are mainly related to certain soil threats. For example urbanization or land take primarily cause soil sealing, soil compaction and decline of organic matter. Industrial uses primarily result in soil sealing and contamination. Agriculture could cause almost all soil threats, even soil contamination by the use of pesticides, manure or fertilizer.

Moreover the kind of use of soil in a specific area must be considered when the needed level of protection is to be determined. This is particularly true with regard to soil contamination. To provide an extreme example: it is obvious that within an industrial site of an oil manufacturing company, a higher level of contamination is acceptable than on a playground of a Kindergarten.

Finally a more scientific issue: soils differ significantly with regard to their type, components, structure. There are an enormous number of different types of soils. Thus each soil type carries its own characteristics and is differently valuable for and vulnerable to different uses.

3.2 Prerequisites of Good Soil Governance

Good governance has to provide solutions for specific challenges. With regard to a good soil governance this means that regulation must first aim to specifically address the various drivers of the soil degradation. The main drivers are—as already mentioned—agriculture, urbanization, land take and industrial practices. These

---

[20]One exemption is if people act in aquatic settings. For example, fishing or shipping do not necessarily have a direct effect on soils.

[21]For example during war, other causes of soil and land degradations may occur.

[22]EU Commission (2006). Additionally, eutrophication is nowadays discussed as a further threat category.

[23]EU Commission (2012) with further information on the status of degradation with regard to the various threats.

drivers of soil degradation significantly differ with regard to the threats, actors and stakes involved and concerning the scientific knowledge.

It is secondly necessary to identify the relevant threats for each of these drivers and to provide appropriate regulatory concepts for the various threats. For example urbanization is a trigger for soil sealing, soil compaction and thus for decline of organic matter. In essence all these soil threats have to be regulated in order to provide an effective management.

This is obviously still a very rough concept for "good soil governance". There is still a need to go much more into detail. Just to provide one example: modern agriculture requires the use of fertilizer and pesticides. The use of both fertilizer and pesticides may cause contamination and might cause degradation of arable land. In order to ensure effective soil protection and management, environmental quality norms, exposure limits and use conditions for each type of fertilizer or pesticides are to be determined.

Nevertheless it shall be underlined that to ensure an improved management of soils, it is certainly not sufficient to put in place general objectives like "soil integrity" or "soil security". It is necessary to deal with the very specific causes of soil degradation processes. The matrix of drivers and threats is a helpful and necessary starting point.

3.3   Excursus: German Soil Protection Governance

In the following excursus it shall be briefly demonstrated that German soil law protection is actually driver and threat specific.

Germany is one of the few states which have a specific Soil Protection Act.[24] The German Soil Protection Act was adopted in 1998 and therefore 2018 marks its 20th anniversary).Paragraph 1 of the German Soil Protection Act stipulates that soil functions have to be sustainably maintained and/or rehabilitated and that to this end both reactive and precautionary measures have to be taken. The overall objective of the law is thus broad enough to encompass all drivers of soil degradation and all soil threats. However, the core operative provisions of this act focus on the regulation of whether and how to clean up contaminated industrial sites including landfills.[25] The law also foresees some provisions with regard to good agricultural practices.[26] However, these provisions are not mandatory and very vague. By way of example, this relates to the requirement that the agricultural practice should

---

[24]Gesetz zum Schutz vor schädlichen Bodenveränderungen und zur Sanierung von Altlasten (Bundes-Bodenschutzgesetz—BBodSchG), 17.3.1998.

[25]§ 4 and 10 of the Soil Protection Act.

[26]§ 17 of the Soil Protection Act.

consider the site specific conditions of the soil. The Soil Protection Act's focus is thus mainly on the rehabilitation of contamination of soils by industrial uses.[27]

As Germany is a very industrialized country, there is also a need to avoid future contamination of soil by ongoing or future industrial activities. Thus the German Act on Immission Control[28] requires an *ex-ante* permission for industrial sites. The permission to build industrial sites may only be granted if detrimental effects on the environment, including soils, are prevented. With regard to the maintenance of soil functions, it is required that the operator—before production starts—examines and documents the status of soils on the industrial site and that the operator after the closure of the site—if significant negative effects have been caused—rehabilitates the sites to the "original" status.[29] The provision clearly sets an incentive to avoid negative effects on soils caused by new industrial uses.

Negative effects on soils by urbanization are—again—dealt with in a specific piece of legislation. The German law on town planning and buildings[30] requires that soils should be managed economically and careful. The act also demands town planning processes by which the various concerns—i.e. need of accommodation and infrastructure, space for industrial uses, protection of the environment, social security—should be taken into account and the potential conflicts between the various concerns could be settled. The act stipulates that if a town planning decisions causes negative effects on nature, including soils, these negative effects have to be compensated.[31] This provision is obviously of eminent importance for nature conservation although it also has some drawbacks in itself.[32] Concerning soils, the provision cannot thoroughly be implemented because a complete assessment of soils is still lacking in Germany.

Agriculture could be the cause of several soil threads. Contamination could result from the input of manure, fertilizer or pesticides. For all these categories of substances specific pieces of legislation are in place in Germany, setting i.e. either emission limit values, quantitative use limits or environmental quality standards.[33] For almost all other—not substance related—soil threats caused by agriculture, such as soil erosion, decline in organic matter, decline in biodiversity or salinization the main trigger for compliance with certain standards is the EU funding system.[34] Funds[35] are only be granted if certain environmental standards are abided by.

---

[27]This is of major importance in Germany due to its high population density. Settlements often are above or close to contaminated grounds.

[28]Gesetz zum Schutz vor schädlichen Umwelteinwirkungen durch Luftverunreinigungen, Geräusche, Erschütterungen und ähnliche Vorgänge (BImSchG), 15.3.1974.

[29]§ 5 IV BImSchG.

[30]Baugesetzbuch (BauGB), 23.6.1960.

[31]§ 1a III BauGB.

[32]Bodle (2017), pp. 297–299.

[33]Möckel et al. (2014), p. 215.

[34]Möckel et al. (2014), p. 473.

[35]The funds make about the half of the farmer's income.

Land take or infrastructure such as streets, airports or railways is dealt with by further pieces of legislation. Infrastructural measures usually require an ex-ante permission. The competent authority has—before granting the permission—to consider all relevant concerns, amongst others also soil (management?). Similar to urbanization, it is mandatory to offset detrimental effects on nature caused by infrastructural measures. Again, concerning the offset of soil degradation, the problem is that a complete assessment of soils is still pending.

In summary, it has been showed that German soil protection governance incorporates drivers of soil degradation and addresses specific threats. The provisions are spread over many pieces of legislation and could thus be described as fragmented.

A core piece of soil legislation is thus missing. There is one exemption: The Soil Protection Act stipulates precautionary limit values with regard to contamination. These precautionary limit values have to be considered by all other laws dealing with soil contamination for example the German Act on Immission Control or the Pesticide Act. With regard to soil contamination, the German Soil Protection Act could therefore be regarded as a kind of a framework regulation. For all other soil threads however, such a framework regulation is missing.

It could be therefore concluded that the essential weaknesses of German soil Governance does not necessarily derive from the so-called fragmentation. For example, an essential shortcoming for the implementation of the compensation obligations in case of urbanization or land take is not cause by the fragmentation but by the fact that soils are not thoroughly assessed.

3.4   Land Degradation Neutrality: How to Implement It?

Land degradation neutrality (LDN) is an objective set by the 2030 Agenda of Sustainable Development. UNCCD has approved a definition of LDN in 2015 which could be regarded as the common understanding of states although not being legally binding:

> Land degradation neutrality is a state whereby the amount and quality of land resources necessary to support ecosystem functions and services and enhance food security remain stable or increase within specified temporal and spatial scales and ecosystems.[36]

Soils are an essential element of land. The LDN objective also applicable to soils.

The term "neutrality" or the words "remain stable or increase" express that the balance of future degradation processes which are unavoidable and future restoration or rehabilitation measures have to be equalized or in simple words "zero".[37]

Good soil governance would have to also comply with this objective, as the LDN objective has been internationally agreed—at least politically. But how to implement

---

[36]UNCCD (2015).

[37]Ehlers (2016) and Minelli et al. (2016).

it? UNCCD has developed the so-called "conceptual framework" which also provides recommendations on the necessary three categories of actions to achieve LDN. The "conceptual framework" uses the term "The LDN response hierarchy".[38] The necessary three categories of actions to achieve LDN are "avoid", "reduce" and "reverse".[39] Avoid is intended that degradation is prevented, in other words, no degradation occurs Reduce means that the negative effects of an activity for soils are minimized to the level possible. "Reverse" should include measure to restore or to rehabilitate degradation processes. Restoration in the view of UNCCD seeks to "re-establish the pre-existing biotic integrity" and rehabilitation aims to "reinstate the ecosystem functionality with the focus on provision of goods and services".[40] In addition the "conceptual framework" sees a need for "a pro-active focus on planning" by which "anticipated losses" should be counterbalanced by "planned gains".[41]

The "conceptual framework" is a very helpful tool in order to promote the implementation of LDN as it provides a plausible structuring of ways and means of its implementation. However, not all questions have to be solved in application of the "conceptual framework". As the LDN objective was only approved in 2015, the already existing legislation has to be evaluated and assessed whether it complies with the requirements of the conceptual framework. Thus states will have to analyze their national legislation whether it provides for suitable mechanism to achieve LDN.

Such an analytical assessment requires the following steps. First it has to be analyzed whether the current legislation provides measures to "avoid" and "reduce" land/soil degradation.[42] Such an analysis could not be undertaken generally but has to be driver and threat specific. The question actually is whether and to which extent degradation is excluded for a certain threat of a driver of soil/land degradation, for example for the operation of an industrial site. If the analysis concludes that for this issue no degradation could occur, no further actions are needed. If this is not the case, the analysis must go on to check whether there are legal provisions in place to ensure that the anticipated losses are to be compensated on a project level. For example in case of the industrial site it has to be analyzed whether the national provisions require that the contamination which will be caused by the industrial activity has to be offset by compensation measures. If this is the case land degradation neutrality is achieved. If not, it has to be analyzed whether additional planning instruments are in place to

---

[38]In addition many other matters have to be clarified, defined and arranged, i.a. the following points: A baseline has to be defined, the status of soil/land have to be assessed, indicators for the assessment of the status, of land degradation and of restoration/rehabilitation actions as well as monitoring requirements have to be defined. Further information see Ehlers (2016), Minelli et al. (2016) and Cowie (2018).

[39]Cowie (2018), p. 30.

[40]Cowie (2018), p. 31.

[41]Cowie (2018), p. 31.

[42]To completely avoid soil/land degradation it is often necessary to prohibit a human activity, because most human activities cause—at least to a certain extent—degradation of soils. Thus it might be reasonable to see "avoid" and "reduce" as one category.

ensure, that degradation not being offset for a specific project are compensated by other means.

It is crucial to understand that probably most if not all national legal system do not foresee a complete compensation scheme for all degradation processes taking into account the various drivers and the soil threats. Thus an overarching planning instrument is certainly needed to identify the major gaps on the project level and to arrange for sufficient restoration or rehabilitation projects.

Taking such an analytical approach would allow states to evaluate whether their current national legislation implements the LDN objective. Thereby the existing gaps could be identified and solved—if there is political will to do so.

4 Conceptual Thoughts on International Soil Governance

In the previous section it has been explained what particularities have to be observed with regard to a good soil governance and how good soil governance has to be designed.

In the following section, some conceptual thoughts will be presented with respect to future international soil governance. To this end, it will first be shown that current international law is fragmented and does not oblige to specific soil related measures. The "regulatory impact" of existing international law with regard to soil protection is thus very limited.

Based on this conclusion and taking as given that states would seek an effective international soil regulation, it will be discussed which conceptual approach of international soil governance would be suitable. Finally an idea will be presented on how such a new soil treaty could be operationalized.

4.1 International Treaty Law: Fragmented and No Specific Measures

First of all, it has to be said that there is no international treaty specifically dealing with soils.[43] Thus no international treaty sets a general framework for sustainable soil management or acts like an anchor for this topic.

The UN Convention to Combat Desertification (UNCCD),[44] the Convention on Biological Diversity (CBD)[45] and the UN Framework Convention on climate

---

[43]Ginzky (2016), p. 11.

[44]The Text of UNCCD could be found under: www.unccd.int.

[45]The Text of CBD could be found under: www.cbd.int.

change (UNFCCC)[46] contain provisions relevant to the sustainable management of soils. All three treaties apply almost universally.[47] However, each of the three treaties has a specific focus and is thus *per se* limited. Moreover no treaty contains soil specific requirements. Soils are therefore only dealt with in a quite general fashion. The following provides a short reasoning with regard to these three regimes.

The UNCCD is the most relevant regime for the management of soil as it expressively refers to "land". However, the UNCCD—formally taken—only deals with the so called dry-lands. Dry-lands taking the legal definition of UNCCD occupy about 40% of the terrestrial surface.[48] Thus, the UNCCD is formally only applicable to less than the half of the terrestrial surface. Moreover, the UNCCD does not contain any soil-specific actions. Countries affected by land degradation are only obliged to implement National Action Programs.[49] No further requirements were agreed although the UNCCD has established itself as the leading regime for the implementation of the objective of "land degradation neutral world" (LDN), by various decisions taken by the Conference of the Parties in 2015.[50]

The scope of application of the CBD could, in general, encompass all soil functions.[51] However, the CBD has only put in place some programs which might have a positive benefit for soils, like the program on agricultural biodiversity. The CBD does not however foresee specific measures with regard to soils.[52]

The prime focus of the UNFCCC are climate change issues. With regard to soils, the UNFCCC provisions on accounting, maintaining and restoring sinks and reservoirs determine a general legal framework for action. The UNFCCC does not however establish specific requirements concerning the management of soils.[53] The Paris Agreement which was signed in December 2015 has set ambitious quantitative targets, which is to hold the increase of global average temperature "well below 2 degree".[54] The Paris Agreement was seen as a major progress in international climate governance.[55] An ad hoc Working Group was established to provide guidelines for the accounting of "nationally determined contributions".[56] It

---

[46]The Text of UNFCCC could be found under: www.unfccc.int.

[47]"Almost", because the United States of America are not a Party to CBD.

[48]Boer and Hannam (2003), p. 153. The determination of whether an area falls within the scope of application of UNCCD depends on the "ratio of annual precipitation to potential evapotranspiration" (Article 1 (g) UNCCD).

[49]Hannam and Boer (2002), pp. 62–63.

[50]Boer et al. (2016), p. 62. See also Wunder et al. (2017).

[51]Ginzky (2016), p. 18. Hannam and Boer (2002), pp. 63–64.

[52]Wolff and Kaphengst (2016).

[53]Boer et al. (2016), p. 58. See also Streck and Gay (2016).

[54]Article 2.1 (a) Paris Agreement.

[55]Boer et al. (2016), p. 59.

[56]Paris Decision, Para 31.3.

has to be determined whether the guidelines will provide requirements for soil and land management.[57]

To sum up, soil related provisions could be found in various treaties such as the UNCCD, the CBD and the UNFCCC. There is no overarching soil treaty which sets a general framework or coordinate the various actions.[58] Neither the UNCCD, nor the CBD or the UNFCCC foresee soil specific provisions. The factual level of regulatory effects on soil conservation is thus very limited.

4.2   General Concepts

Any time the need of additional international rules for soils is mentioned, the usual reaction for the time being is that there is no appetite within the international community for new treaties or organizations. This simple and at the first glance convincing statement is actually wrong—at least in its generality. The following should quickly underline that in the last years several international treaties were substantially amended or even newly adopted. In 2013, for example, the Contracting Parties of the 1996 Protocol to the Convention on the Prevention of Marine Pollution by Dumping of Wastes and other Matter,[59] adopted a major—legally binding—amendment to the London Protocol concerning marine geo-engineering.[60] The Minamata- Convention on Mercury[61] and the Paris Agreement were adopted in 2013 and in 2015. Both of them are therefore newly adopted international treaties.

There have been also several proposals for proposals for additional instruments on soil governance which have been submitted by experts. First the IUCN draft Protocol on security and sustainable use of soils has to be mentioned.[62] Second: an expert group which was established after the first Global Soil Week in Berlin 2013 provided the idea of a "thematic annex" to UNCCD on soil protection.[63] However, no negotiation process concerning an international regime has been set up so far. Even the UNCCD although establishing itself as the international lead organization

---

[57]See Fee (2018).

[58]Hannam and Boer (2002), p. 62.

[59]Further information under http://www.imo.org/en/OurWork/Environment/LCLP/Pages/default.aspx.

[60]Section 4 and Annex 4 of the *Report of the 35th Consultative Meeting of Contracting Parties to the London Convention and 8th Meeting of Contracting Parties to the London Protocol*, LC 35/15, 21 October 2013.

   For an assessment of the amendment see Ginzky and Frost (2014).

[61]Further information under: http://www.mercuryconvention.org/Convention/Text/tabid/3426/language/en-US/Default.aspx.

[62]IUCN Draft Protocol on security and sustainable use of soils, Sept 2009, by IUCN Commission on Environmental Law Specialist Group on Sustainable Use of Soils and Desertification.

[63]A discussion paper is available under: http://www.iass-potsdam.de/sites/default/files/files/gsw_leg_discussion_paper_options_unccd_endg_0.pdf.

on LDN focusses its current work on pilot projects in more than 100 countries with the aim of "voluntary target setting" for LDN.[64] Parties to the UNCCD have not initiated a debate to amend UNCCD for soil issues.[65]

Nevertheless: political perception may change. To illustrate this, the case of marine geo-engineering under the London Protocol could be used to illustrate this: initially, only a non-legally binding resolution on ocean fertilization[66] in particular was adopted by the Contracting Parties in 2008. It was then arduously debated whether a legally binding regime would be suitable.[67] Finally in 2013, the Contracting Parties adopted the legally binding amendment as mentioned above. It is interesting to see that one core reason for the willingness of states to adopt a legally binding instrument was the fact that several most controversial fertilization projects had been conducted between 2008 and 2013.[68]

There are good arguments that more must be done on international level to achieve a better protection of soils worldwide.[69] For the following observations on conceptual approaches it should be assumed that the global community is willing to work on this "more" out of international soil governance.

4.2.1 Keeping the status quo

First option of future international soil governance would be to mainly keep the status quo.

The main characteristics of the CBD and the UNCCD governance could be described as follows[70]:

- Strong emphasis on national sovereignty of Parties
- Different responsibilities of states
- International financial and technical cooperation[71]
- Awareness of the need for additional scientific information and knowledge
- Implementation by national strategies, programs or plans

The CBD requires in addition the following:

- Application of the precautionary approach

---

[64]Minelli et al. (2016), pp. 87–88. More information under: https://www2.unccd.int/actions/ldn-target-setting-programme.

[65]Mastrojeni (2018).

[66]The resolution could be found under: http://www.imo.org/en/OurWork/Environment/LCLP/EmergingIssues/geoengineering/Pages/default.aspx.

[67]Ginzky (2010) and Markus and Ginzky (2011), p. 479.

[68]Ginzky and Frost (2014), p. 94.

[69]See the arguments under section 1.

[70]See Bowling et al. (2016), p. 9.

[71]Morgera and Tsioumani (2010), p. 28.

- Importance of in-situ conservation by States applying inter alia Environmental Impact Assessments (Article 8 and 14 CBD)
- Sharing of benefits derived from the exploitation of genetic resources[72]

An Environmental Impact Assessment would be an appropriate tool to minimize the effects on soils. However, it is not required by the UNCCD. The UNCCD has until today not included the obligation to apply a precautionary approach. Such an approach would certainly be a helpful tool to implement an efficient soil management and to achieve Land Degradation Neutrality. The requirement of sharing of benefits is however not relevant for sustainable soil management. Whereas the generic information in the case of the CBD has an economic value beyond its local availability and functionality this is different to soils because the economic value of soils is in principle bound to the use of the soil itself.[73]

The pilot country projects on voluntary target setting for LDN established by UNCCD[74] is an expression of the main characteristics mentioned above. It emphasizes the national sovereignty and the different responsibilities of states, is based on financial cooperation, stresses the need of additional information and is actually implemented at national level.

The conceptual approach of these two treaties is to define a general objective—conservation of biodiversity, combatting desertification—and to put in place methods of international cooperation. The specific measures to achieve these ends are to be determined by national legislation. It is actually to the discretion of states how ambitiously the general obligations of the CBD and the UNCCD are implemented nationally.

Thus, the concrete outcome for soils and for the maintenance, strengthening or rehabilitation of ecological services of soil (=the regulatory impact of the existing international provisions) is not clear as there are no soil specific obligations set by the UNCCD and the CBD which have to be complied with by states.

4.2.2 Coordination Treaty

Another option of future international soil governance would be a "coordination treaty".[75] The fundamental aim of such a treaty would be to coordinate the existing regimes with regard to their soil related objectives, principle, instruments, mechanisms and procedures. The objective would be to provide for a better interaction of the existing regimes by coordinating the actions of the international organizations as well as the obligations of the Contracting Parties.[76]

---

[72]Text of the respective protocol under: https://www.cbd.int/abs/doc/protocol/nagoya-protocol-en.pdf.

[73]Ginzky (2017), p. 442.

[74]Minelli et al. (2016), pp. 87–88.

[75]Eikermann (2017), p. 428.

[76]Eikermann (2017), pp. 428–429.

In fact, a better coordination in particular between the UNCCD and the UNFCCC is certainly required as soil is very important for climate change mitigation because almost all actions to increase CO_2-concentrations in soil have positive effects on soil quality. Thus to improve soil quality is in the interest of UNFCCC and UNCCD. Coordinated actions would be more than reasonable.

The proposal of an international "coordination treaty" however raises many questions which have neither been fully discussed nor solved. One issue concerns the question how coordination, in case that it is desired, should be achieved. Would it be by decisions taken under the umbrella of the coordination treaty which then would be mandatory for the other regimes (UNCCD, CBD, UNFCCC) This would require a kind of a hierarchy between the existing regimes (UNCCD, CBD, UNFCCC) and the new coordination treaty which is hardly imaginable. To put in place such a hierarchy would also require that all regimes have been ratified by the same states.[77]

The coordination could also be achieved by procedural requirements. The coordination treaty would then set up certain procedures to arrange for a coordinated approach between the different regimes. However, it would have to be explained why for such procedural arrangements a new international treaty would be necessary. Such procedural arrangements could also be regulated or framed by Memoranda of Understanding (MoU) between the various international regimes (UNCCD, CBD, UNFCCC). A MoU is a kind of a contract between the different organizations which only bounds the parties to this contract. MoU usually are used to agree on organizational arrangements. To agree on a MoU is much easier to have than an international coordination treaty as it does not have to be ratified by states.

4.2.3 More Specific Obligations

The third option of future international soil governance is to put in place more specific obligations by new international provisions. The basic idea is that states are clearly obliged to enact certain measures which should have then identifiable benefits for soils. This conceptual approach is inter alia used for the London Protocol. The overall objective of London Protocol is to protect the marine environment from negative effects caused by dumping, meaning the disposal of waste or other matter into the oceans.[78] The regulatory approach is simple and clear. In general all dumping is prohibited by LP. States are only allowed to consider dumping of eight waste categories which are expressively mentioned in Annex 1 of the London Protocol. The disposal of waste of one of these categories requires

---

[77]Treaties are only mandatory for their Parties. Therefore it would be required to have more or less the same membership.

[78]See: http://www.imo.org/en/OurWork/Environment/LCLP/EmergingIssues/geoengineering/Pages/default.aspx.

a permit. Before granting a permit states have to assess and to minimize the effects on the marine environment.[79]

Something comparable could also be established for soils. As mentioned in Sect. 3 good soil governance requires to be a driver and threat specific. International law would have to determine the essential measures for the various drivers and the specific soil threats. For example: concerning the driver urbanization, international law could entail an obligation to establish norms, standards and procedures for town planning. Concerning contamination by industrial sites, an international agreement could oblige to carry out an ex-ante assessment of the effects to soil, to minimize the effects and to require a permit before starting the operation.[80]

In this context it should be mentioned that a certain level of flexibility in implementing these obligations should be conceded as the political, economic and technical realities differ among states worldwide. The political priorities should be determined by each society.[81] Therefore, the international obligations which requires specific measures to be taken for certain drivers should be limited to the most essential ones and provide substantial room how they are implemented on national level.

4.2.4   Summary

There are different options on how to develop the existing international soil governance. It has been shown that the existing international soil governance by the UNCCD, the CBD and the UNFCCC is fragmented and does not entail specific soil related obligations. All presented concepts have advantages on their own. The first option "keeping the status quo" foresees a regulatory concept which is based on a general objective and which arranges for a good international cooperation—with all the weaknesses mentioned. This concept emphasizes the sovereignty of states and assigns the states a broad discretion how to implement the general objective.

The concept of a coordination treaty is the answer to the required better streamlining of international work which is also of extreme importance. Memoranda of Understanding (MoU) would be an appropriate instrument to achieve more integrated and concise actions. Such MoU should be negotiated between the various regimes (UNCCD, CBD and UNFCCC) to better coordinate the existing regimes.

To entail international obligations for more specific measures could supplement the existing regime. The advantage would be to have more predictable outcomes for the benefits of soils. These additional obligations should be limited to the most important aspects and should provide for sufficient flexibility with regard to national characteristics and priorities.

---

[79]Article 4 of London Protocol.

[80]Ginzky (2016), pp. 27–30.

[81]Boer et al. (2016), p. 69.

4.3   Regime Framing

International treaties could be very differently conceptualized. One concept is to include all relevant provisions in one piece of legislation. The IUCN proposal for a soil treaty is based on this concept.[82]

Another concept is to have a convention with an enabling clause for protocols. The CBD and the UNFCCC are structured this way. The protocols are however independent international treaties and do not form an integral part of their framework conventions and therefore have to be ratified individually.

Thirdly, a convention could entail fundamental provisions like objectives, principles, obligations and procedural rules. In Annexes to the convention, specific aspects which are relevant in a specific field could be regulated. The already mentioned London Protocol follows this approach. The objective, principles and general obligations are regulated in the main body of the instruments. Annex 1 foresees the waste categories which may be considered for dumping. Annex 2 entails the criteria for the assessment of dumping activities. Amendments to the Annexes "shall be adopted by a two-thirds majority vote of the Contracting Parties present and voting at a Meeting of Contracting Parties" pursuant to Article 22 Paragraph 2 London Protocol. Amendments enter into force after 100 days after its adoption for all Contracting Parties except for those which submit a declaration of non-acceptance.

The substantial advantages of this concept would be that the amendments to annexes do not have to be ratified. Amendments could be adopted and enter into force by vote of Contracting Parties. Each state would still have the option to opt-out for amendments which the state does not deem to be suitable or appropriate for their national situation.

For international soil governance, the envisaged convention could contain more general provisions. In annexes, provisions for the specific drivers and for the relevant threat could be included. Such a concept would allow to agree on a general political outline and to develop the specifics with respect to drivers and threats over time. Moreover, this concept would allow to review and update the driver and threat specific provisions periodically and to adjust them to increasing knowledge.

5   Conclusions

Good governance for sustainable management of soils—such a demanding task.

The chapter has shown that sustainable management of soils is a global challenge because soils provide very important ecological services. With the sustainability agenda of 2015 and other international agreements this statement was also endorsed by world community.

---

[82]IUCN Draft Protocol on security and sustainable use of soils, Sept 2009, by IUCN Commission on Environmental Law Specialist Group on Sustainable Use of Soils and Desertification.

Good soil governance has to be a driver and thread specific in order to be effective. Such a conceptual approach is key to get real regulatory impacts. The analysis of German soil governance has shown that fragmentation of provisions among different legislative instruments is *per se* not a drawback as long as the provisions entail suitable requirements.

To implement the objective of LDN, a detailed analysis of current legislation is required—taking the response hierarchy "avoid" and "reduce", "reverse" and "planning instruments" as structuring criteria.

The existing international law does currently not entail an effective regulation of soils. It does not encompass specifically dedicated obligations that are aim expressly to ensure sustainable soil management. Good governance on international level requires instruments for cooperation, for coordination between the various regimes (UNCCD, CBD and UNFCCC) and should also entail obligations for driver and threat specific actions. Instruments for international cooperation are already entailed in the existing regimes. Coordination could be implemented by MoU. A specific coordination treaty seems to be not the most appropriate tool. Obligations for driver and threat specific actions would have to be embedded in existing treaties or in a new treaty.

If a new treaty would be the political choice, a framework treaty with annexes in which the specific obligations could be regulated seems to be a practical concept. The core provisions could be entailed in the framework treaty whereas the driver specific regulations could be included in the annexes. The advantage would be that first the international governance would be driver/threat specific and second the annexes could be amended without being ratified by all states.

Food for thoughts—nothing more. Currently, the main international activity is the target setting program run by the UNCCD. In this course there may be a chance to develop "best practices" models for the implementation of the UNCCD objectives and the LDN objective. These models could then be taken by other states to amend their legislation.

The aim of this chapter was twofold. First to provide some conceptual thoughts on good soil governance in general which might be helpful to work out "best practices" models. And second: to provide some thoughts concerning future international soil governance in case a window of opportunity may open. It is a tragedy that such windows of opportunity mostly result from crisis—like the food crisis with regard to the increased awareness on soils.

References

Bodle R (2017) Implementing land degradation neutrality at national level: legal instruments in Germany. Int Yearb Soil Law Policy 2:287–308
Boer B (2014) Land degradation as a common concern of humankind. In: Lenzerini F, Vrdoljak F (eds) International law for common goods, pp 289–307
Boer B, Hannam I (2003) Legal aspects of the sustainable soils. Rev Eur Com Int Environ Law 12(2):149–163

Boer B, Ginzky H, Heuser I (2016) International soil protection law: history, concepts and latest developments. Int Yearb Soil Law Policy 1:49–72

Bowling C, Pierson E, Ratté S (2016) The common concern of humankind: a potential framework for a new international legally binding instrument on the conservation and sustainable use of marine biological diversity in the high seas. White Paper, 1–15, under: http://www.un.org/depts/los/biodiversity/prepcom_files/BowlingPiersonandRatte_Common_Concern.pdf

Cowie AL, Orr BJ, Castillo Sanchez VM, Chasek P, Crossman ND, Erlewein A, Louwagie G, Maron M, Metternicht GI, Minelli S, Tngberg AE, Walter S, Welton S (2018) Land in balance: the scientific conceptual framework for land degradation neutrality. Environ Sci Policy 79:25–35

Ehlers K (2016) Chances and challenges in using the sustainable development goals as a new instrument for global action against soil degradation. Int Yearb Soil Law Policy 1:73–84

Eikermann A (2014) Der Wald im internationalen Recht: Defizite, Regelungsoptionenund Mindestanforderungen. Rechtsgutachten im Auftrag des Bundesamtes für Naturschutz

Eikermann A (2017) International forest regulation: model of international soil governance. Int Yearb Soil Law Policy 2:413–432

EU Commission (2006) Communication: Thematic Strategy for Soil Protection, COM (2006) 231 final

EU Commission (2012) Report: The implementation of the Soil Thematic Strategy and ongoing activities, COM (2012) 46 final

Fee E (2018) Implementing the Paris Climate Agreement: risks & opportunities for sustainable land use. Int Yearb Soil Law Policy 3:249–269

Flasbarth J (2016) Soils need international governance: a European perspective for the first volume of the International Yearbook of Soil Law and Policy. Int Yearb Soil Law Policy 1:15–20

Ginzky H (2010) Ocean fertilization as climate change mitigation measure – considerations under international law. J Eur Environ Plann Law:57–78

Ginzky H (2016) Bodenschutz weltweit – Konzeptionelle Überlegungen für ein internationales Regime. Handbuch Boden:1–32

Ginzky H (2017) The sustainable management of soils as a common concern of humankind: how to implement it? Int Yearb Soil Law Policy 2:433–450

Ginzky H, Frost R (2014) Marine geo-engineering: legally binding regulations under the London Protokoll. Carbon Clim Law Rev 2014:82–96

Hannam I, Boer B (2002) Legal and institutional frameworks for sustainable use of soils. IUCN Environmental Policy and Law Paper No. 45

Heuser IL (2017) Development of soil awareness in Europe and other regions: historical and ethical reflections about European and (international) soil protection law. Int Yearb Soil Law Policy 2:451–474

Linz F, Lobos I (2016) Boden und Land in der internationalen Nachhaltigkeitspolitik – von der globalen Agenda zur lokalen Umsetzung. Zeitschrift für Umweltrecht:195–199

Markus T, Ginzky H (2011) The regulation of climate engineering – conceptual considerations based on the example ocean fertilization. Carbon Clim Law Rev:477–491

Mastrojeni G (2018) UNCCD 13: from awareness to action in a complex world. Int Yearb Soil Law Policy 3:229–247

Minelli S, Erlewein A, Castillo V (2016) Land degradation neutrality and the UNCCD: from political vision to measurable targets. Int Yearb Soil Law Policy 1:85–104

Möckel S, Köck W, Rutz C, Schramek J (2014) Rechtliche und andere Instrumente für vermehrten Umweltschutz in der Landwirtschaft. Berlin

Morgera E, Tsioumani E (2010) Yesterday, today and tomorrow: looking afresh at the convention on biological diversity. Yearb Int Environ Law:3–40

Streck C, Gay A (2016) The role of soils in international climate change policy. Int Yearb Soil Law Policy 1:105–128

UNCCD (2015) Report of the Conference of the Parties on its twelfth session, held in Ankara form 12 to 23 October 2015. Part two. ICCD/COP(12)/20/Add.1 http://www.unccd.int/Lists/OfficialDocuments/cop12/20addd1eng.pdf

United Nations (2012) General Assembly, Resolution adopted by the General Assembly on 11 September 2012, 66/288, The Future We Want

United Nations (2015) General Assembly, Seventieth Session, No. 11688, Agenda items 15 and 116, Resolution adopted by the General Assembly on 25 September 2015, 'Transforming our world: the 2030 Agenda for sustainable development', A/RES/70/1, p 1

Wolff F, Kaphengst T (2016) The UN Convention on biological diversity and soils: status and future options. Int Yearb Soil Law Policy 1:129–148

Wunder S, Kaphengst T, Freilih-Larsen A (2017) Implementing land degradation neutrality (SDG 15.3) at national level: general approach, indicator selection and experiences from Germany. Int Yearb Soil Law Policy

Legal and Regulatory Framework for the Agriculture Sector in Uganda

Emmanuel Kasimbazi

1 Introduction

Agriculture can be described as the science of cultivating the soil, harvesting crops and raising livestock and also the science or art of the production of plants and animals useful to man and in varying degrees, the preparation of such products for man's use and their disposal.[1] The agricultural sector has been and continues to be the most important sector in Uganda's economy in terms of food and nutritional security, employment, income, raw materials for industry and exports to regional and international markets. It employs 82% of the workforce, accounts for 90% of export earnings, and provides 44% of GDP.[2] Moreover, 2.5 million farmers in Uganda with smallholdings and scattered large commercial farms provide the country's staple food requirements.

According to Trading Economics publication 2018, GDP from Agriculture in Uganda increased to 4208.28 billion Uganda Shillings in the third quarter of 2018 from 3285.33 billion Uganda Shillings in the second quarter of 2018. GDP from the agriculture sector in Uganda averaged 2901.67 billion Uganda Shillings from 2008 until 2018, reaching an all-time high of 4208.28 billion Uganda Shillings in the third quarter of 2018.[3]

Uganda is able to rely on agriculture due to the country's excellent access to water resources, fertile soils, and (relative to many other African nations) its regular

---

[1]Black (1990).

[2]Negm and Abdel-Fattah (2017).

[3]Trading Economies (2019).

E. Kasimbazi (✉)
School of Law, Makerere University, Kampala, Uganda
e-mail: ekasimbazi@law.mak.ac.ug

© Springer Nature Switzerland AG 2020
H. Yahyah et al. (eds.), *Legal Instruments for Sustainable Soil Management in Africa*, International Yearbook of Soil Law and Policy,
https://doi.org/10.1007/978-3-030-36004-7_4

rainfall, although it does still suffer from intermittent droughts such as in 1993–94.[4] Uganda's favorable soil conditions and climate have contributed to the country's agricultural success. Most areas of Uganda receive enough rain to support subsistence agriculture. Some areas of the southeast and southwest can receive average rain of more than 150 millimeters per month and in the north, there is often a short dry season in December and January. The temperatures vary only a few degrees above or below 20 °C but are moderated by differences in altitude. These conditions have allowed continuous cultivation in the south but only annual cropping in the north, and the driest northeastern corner of the country has supported only pastoralism. Although population growth has created pressure for land in a few areas, land shortages have been rare, and only about one-third of the estimated area of arable land is under cultivation.[5] These favorable conditions have led to a steady increase in the total agricultural production; the total number of livestock (cattle, sheep, goats, pigs and poultry) increased from 72,175,000 in 2013 to 83,140,000 in 2015 and 86,355,000 in 2017.[6] Taking into account the fact that Uganda mostly exports agricultural products (80% of total exports), Exports in Uganda increased to 333.85 US\$ Million in October from 300.12 US\$ Million in September of 2018. Exports in Uganda averaged 129.61 US\$ Million from 1993 until 2018, reaching an all-time high of 351.73 US\$ Million in May of 2017 and a record low of 12.39 US\$ Million in July of 1993.[7]

Uganda's key agricultural products can be divided into cash crops, food crops and horticultural produce. The most important cash crops are coffee, tea, cotton, tobacco and cocoa. Uganda was second only to Kenya as Africa's largest producer of tea, exporting US\$17.06 million of tea in 1996 and US\$39 million by 1998.[8] Unmanufactured tobacco exports provided US\$9.5 million in 1998, over 25% more than in 1996.[9]

The export of cocoa beans hit a recent high in 1996 with US\$1.07 million in export receipts, but this declined to \$0.87 million in 1998.[10] The primary food crops, mainly for domestic consumption, include plantains, cassava, maize, millet, and sorghum. Total cereal production was 1.76 million metric tons in 1998, which provided US\$17.82 million of exports in 1998. This gain was in part negated as imports of cereals were \$30.9 million in the same year.[11] The more recent development of cultivating horticultural produce includes fresh flowers, chilies, vanilla, asparagus and medicinal plants.[12] It is unclear how well horticultural production

---

[4]Nations Encyclopedia (2018).
[5]Byrnes (1992).
[6]UBOS (2018).
[7]Trading Economies (2019).
[8]Trading Economies (2019).
[9]Trading Economies (2019).
[10]Nations Encyclopedia (2011).
[11]Nations Encyclopedia (2011).
[12]Nations Encyclopedia (2011).

will prosper but it does indicate the economy's potential diversity. The fact that vanilla production is the third largest in Africa, providing US$930,000 in export receipts in 1998, is a success in itself.[13]

The economy of Northeastern Uganda is dominated by pastoralism (cattle farming).[14] Although agricultural production is apparent in some areas, this is normally a mixture known as "agro-pastoralism" (integrated cattle and crop farming).[15] It should be noted that pastoralism declined due to the constant cattle raids by guerrilla groups such as the Lord's Resistance Army, as well as government and aid agency interventions which encourage the fencing off of land to discourage the traditional free-roaming of cattle.[16]

The Ministry of Agriculture, Animal Industry and Fisheries (MAAIF) is the key ministry that oversees the agricultural sector in Uganda. It formulates, reviews and implements national policies, plans, strategies, regulations and standards, and enforces laws, regulations and standards along the value chain of crops, livestock and fisheries.[17] There are several policies, laws and regulations that provide a framework for investing in agriculture. The main purpose of this chapter is to analyze how the key legal issues in the agricultural sector are regulated. These issues include regulation of land for agriculture purposes, soil governance, livestock management, crop production and management, regulation of water for agriculture, pesticide control, fertilizer management and regulation of equipment and machinery as well as taxation.

2 Analysis of the Policy, Legal and Regulatory Framework for the Agricultural Sector

The key legal issues in the Agricultural sector stated above are contained in the Constitution, Acts of Parliament, policies, regulations, strategies and plans.

2.1 Regulation of Land for Agricultural Purposes

Land is the most important household asset for households that depend on agriculture for their livelihoods. Therefore, access to land is a basic requirement for farming and control over land. The regulation of land for agricultural purposes is contained in

---

[13]Nations Encyclopedia (2011).

[14]Nations Encyclopedia (2011).

[15]Nations Encyclopedia (2011).

[16]Nations Encyclopedia (2011).

[17]Government of the Republic of Uganda, Ministry of Agriculture, Animal Industry and Fisheries (2013).

the Constitution and other laws. The 1995 Constitution of Uganda provides for environmental protection and conservation. The Constitution also sets out National Objectives and Directive Principles of State Policy under which Objective XIII provides that the state shall protect important natural resources. Objective XXVII provides for sustainable environmental management and Article 39 provides that every Ugandan citizen has a right to a clean and healthy environment. The Constitution further makes provisions for land ownership. Article 237 provides that all land belongs to the people of Uganda and shall be owned in accordance with the land tenure systems; customary, freehold, mailo and leasehold.

In addition, under the Constitution, Article 237(2)(b) provides that the government or a local government shall hold in trust for the people and protect natural lakes, rivers, wetlands, forest reserves, game reserves, national parks and any land for ecological and tourism purposes for the common good of all citizens. These provisions imply that all the people of Uganda have a stake in the management of natural resources and the resources thereon. Specifically, Article 189 read together with Schedule 6 make the Central Government responsible for the Agricultural Policy. This implies that farming on any land should take into account the environment, people's interests in land and development principles prescribed under the Constitution.

The Land Act Cap 227 is the main law on land management and provides for the tenure, ownership and management of land for agriculture. According to Sections 2 and 3 of the Act, the ownership of land is vested in the citizens of Uganda and land can be owned under customary, freehold, mailo and leasehold tenure systems.

The Act promotes land registration[18] (the creation of a customary register to facilitate the registration of customary rights and issuance of customary certificates of ownership) and conversion of customary tenure to freehold tenure,[19] conversion of leasehold into freehold[20] and formation of a communal land association by any group of persons connected with communal ownership and management of land, whether under customary law or otherwise.[21]

It is important to note that land and resource ownership determine use, who benefits and who has rights and responsibilities for the land and its resources. Therefore, to some extent agricultural activities in Uganda are greatly determined by land rights. The Land Act makes provisions for gender equality in land rights. Section 40 requires that before any transaction takes place regarding land on which a family lives and/or derives their sustenance, the spouse and adult dependent children should be consulted. The Act under Section 28 complements the 1995 Constitution by stating that the customary practices that deny women or children use of land are null and void. The local Land Committees set up in each parish are to ensure that these provisions are carried out and that vulnerable groups are protected. Taking into

---

[18]Section 5 of the Land Act.

[19]Section 10 of the Land Act.

[20]Section 29 of the Land Act.

[21]Section 16 of the Land Act.

consideration the fact that women and children provide the greatest labour on farmlands, the protection of their rights of access to land for agriculture is key.

Furthermore, Section 40 of the Act restricts ownership of land by noncitizens. Under this section, noncitizens can only hold land under leasehold tenure for a period not exceeding 99 years. This may restrict foreigners who may want to participate in long term commercial farming. The law does not prescribe which agricultural activities should be carried out under a specific land tenure system and as such, the land tenure systems do not impact on agriculture in any way. It should however be noted that some farmers in Uganda grow crops and rear animals on land owned by someone else. The development of land lord tenancy laws and concepts which provide for the temporary use of land by another, with assurances of being paid rent, make the extensive use of leases within Ugandan agriculture important.

The land tenure system certainly has an influence on agricultural production, but other factors normally operate to a greater degree. Such factors include: the absence of a horizontally integrated economy, the nature of landholding, land fragmentation and poor farming techniques.

The National Land Policy promulgated in 2013 provides a framework for development and use of Uganda's land resources. The Policy has two major objectives:

- To re-orient the land sector in national development by articulating management and co-ordination between the land sector and other productive sectors in the economy; and
- Enhancing the contribution of the land sector to the social and economic development of the country.[22]

The goal of the National Land Policy is to ensure an efficient, equitable, optimal utilisation and management of Uganda's land resources for poverty reduction, wealth creation and overall socio-economic development. Policy Statement 148 provides that government shall regulate the use of land and water resources for agricultural production aligned with a National Agricultural Policy. The strategies of achieving the above policy statement require government to:

- Formulate a national agriculture policy;
- Formulate a national soils policy;
- Promote and ensure viable zonal agricultural production to enhance production, productivity, marketing and agro-processing;
- Make available an updated soil and arable land resource inventory at an appropriate scale;
- Promote farming practices that reduce land degradation and enhance soil quality and productivity;
- Encourage voluntary consolidation of agricultural land holdings to sizes suitable for optimum, productive and sustainable use;

---

[22]Kabanda (2013).

- Plan, use and regulate agricultural activities and other practices that degrade the quality of agricultural land;
- Discourage land fragmentation through education, incentives, laws and byelaws; and
- Promote sustainable use and management of water, soil and land resources.

Despite the existence of the policy, land productivity potential, land capability and land sustainability for agriculture is not well known. This is because there has been very limited research into the land productivity potential and land capability. The research culture in land productivity potential, land capability and sustainability is still lacking. This has contributed greatly to the misuse of soil for agriculture, affected soil quality and soil governance. This makes it nearly impossible to allocate agricultural land to its most optimal and sustainable use. Agricultural zones of production excellence based on production potential and existing comparative advantages, though self-evident are not demarcated.

Poor agricultural practices have resulted into increased land degradation due to soil erosion and soil nutrient depletion,[23] de-forestation, over grazing and water contamination. Some of the poor agricultural practices include: poor techniques of threshing, outdated practices involving shelling, drying, storing,[24] packaging and transporting of agricultural produce. Overpopulation in some areas has resulted in land fragmentation and overuse, affecting land quality, agricultural production and economic development. Land tenure security as it relates to access and ownership remains a major menace for women farmers.

The policy is important for optimal utilisation and management of land resources especially regarding agricultural investments. However, the successful implementation of the national land policy will depend on continued buy-in, support and confidence of all stakeholders. Stakeholders must participate and be constructively engaged at all levels of policy implementation. These include different government departments, development partners, the private sector, civil society organisations, professional bodies, cultural institutions, faith-based organisations and other non-state actors.

2.2   Soil Governance

Soil governance refers to the policies, strategies, and the processes of decision making employed by states and local governments regarding the use of soil. Governing the soil requires international and national collaboration between governments, local authorities, industries, civil society organisations and citizens to ensure implementation of coherent policies that encourage practices and

---

[23]Conserve Energy Future (2018).

[24]The New Vision (2012).

methodologies that regulate usage of the resource to avoid conflict between users to promote sustainable land management. Soil is the primary medium for food production.

Soil governance is directed towards the impacts of soil degradation on food production and conflicts that arise between the need for human settlements and space available for food production.[25] Therefore, effective soil governance is important for the agricultural sector.

2.2.1 Regulatory Framework for Soil Governance

The National Environment Act (NEA) Cap 153

This Act emanated from the National Environment Action Plan (NEAP) aims at providing for the sustainable management of the environment and natural resources. Section 40 (2) of the Act mandates the National Environment Management Authority (NEMA) to issue guidelines and prescribe measures for the sustainable use of hillsides, hilltops and mountainous areas including those relating to appropriate farming methods, carrying capacity of the areas in relation to animal husbandry and measures to curb soil erosion. In order to operationalise the broad measures above, NEMA has issued regulations and standards to guide the sustainable use of environmental resources that are relevant to agricultural production in accordance with Section 107 of the Act.

The National Environment (Minimum Standards for Management of Soil Quality) Regulations, Statutory Instrument Number 59 of 2001

The Regulations establish and prescribe minimum soil quality standards to be maintained for the management of the quality of soil, the criteria and procedure for the measurement and determination of soil quality and guidelines for soil management. The fourth schedule of the Regulations provides that soil conservation is required as a basis for environmentally sound production of food, wood and other commodities based on sustainable use of land, species and ecosystems. Moreover, combination of several conservation practices is recommended and packages should depend on area and crops/livestock/tree species on the land. A responsible person is defined to mean the owner of the land or person residing on, or using the land.

The standards under regulation 12 impose an obligation on every responsible person to comply with the measures and guidelines of soil conservation for the particular topography, drainage and fanning systems prescribed in the Fourth Schedule. Noncompliance is an offence punishable by a fine of not less than one hundred

---

[25]Lal et al. (1989).

and eighty thousand shillings and not more than eighteen million shillings or to imprisonment not exceeding eighteen months, or both.

Regulation 13 empowers the environmental officers to ensure compliance with the standards set out in the schedules to the standards with power to issue improvement notices and to enter onto the land and ensure compliance with the said standards. Non-compliance with the notice is an offence punishable on conviction, by a fine of not less than one hundred and twenty thousand shillings and not exceeding twelve million shillings, or to imprisonment for a term not exceeding twelve months or both. By the standards imposing duties on responsible persons and further empowering environmental officers to ensure compliance with the soil standards, these standards are therefore applicable by land owners since there are environmental officers at every district. However the application of these standards has its challenges as the people need sensitisation on first of all their existence and secondly, their applicability for them to be successfully applied.

The National Environment (Wetlands, Riverbanks and Lakeshores Management) Regulations, 2000

The Regulations provide for the conservation and wise use of wetlands and their resources and facilitate the sustainability and conservation of resources on riverbanks and lakeshore by and for the benefit of the people and community living in the area. Regulation 7 enjoins the district and environment committees with the responsibility of coordinating, monitoring, and advising District Councils on all aspects of wetland resource management. The local environment committee is enjoined with the responsibility of being the implementing organ in conserving and managing wetland resources in its area of jurisdiction. The regulations empower the minister to declare certain wetlands especially protected wetlands of national or international importance. Regulation 11 provides for uses of wetlands where a person desiring to carry out any of the regulated activities listed in the Second schedule or extract in a wetland shall make an application except for activities such as any cultivation where the cultivated area is not more than 25% of the total area of the wetland and fishing using traps, spears and baskets or other method other than weirs.

Regulation 12 prohibits the carrying out of any activity in a wetland without a permit issued by the Executive Director. Every landowner, occupier or user who is adjacent or contiguous with a wetland is enjoined to prevent the degradation or destruction of the wetland and shall maintain the ecological and other functions of the wetland. The regulations obligate any person who intends to introduce or plant any plant whether alien or indigenous on a river bank or lake shore, or introduce any animal or micro-organism, whether alien or indigenous in any river bank or lake shore, to make an application to the executive director. Permits granted under the regulations can be revoked by the Executive Director.

Regulation 27 enjoins an environment officer within whose jurisdiction activities likely to degrade the environment, river banks or lake shores are taking place, to ensure that the communities living near a wetland participate in its conservation and assist environment committees in implementing these Regulations and any other law that protects wetlands. Every land owner or user in whose land a river bank or lake shore is situated is under a duty to prevent and repair degraded river banks and lake shores through agro-forestry; grassing; and control of livestock grazing. Non-compliance with any of the obligations imposed under the regulations is an offence.

It follows therefore that whoever is under a specific obligation under the regulations must perform their respective obligations or risk the sanctions under the same regulations.

The National Environment (Environmental Impact Assessment) Regulations, 1998

The Regulations provide procedures to follow when conducting an environmental impact assessment. Regulation 3 provides for the application of the regulations to all projects included in the Third Schedule to the Act. In the third schedule of the NEA, the projects to be considered for Impact Assessment include agriculture related activities such as large scale agriculture, use of new pesticides, introduction of new crops and animals, and use of fertilizers.

The Regulations prohibit developers from implementing projects for which environmental impact assessment is required under the Act and under the regulations unless the environmental impact assessment has been concluded in accordance with the regulations. Regulation 12(2) imposes an obligation on a developer to hold meetings with the affected communities to explain the project and its effects. Regulation 19 imposes an obligation on the Executive Director of the lead agency to invite the general public to make written comments on the environmental impact assessment. The Executive Director is further enjoined to invite comments of those persons who are most likely to be affected by the proposed project. The Executive Director is enjoined to call for a public hearing under regulation 21 where there is a controversy or where the project may have trans-boundary impacts.

The Executive Director is empowered under Regulation 28 at any time after the issuance of a certificate of approval of the project, to revoke the same where there is non-compliance with the conditions set out in the certificate; where there is a substantial modification of the project implementation or operation which may lead to adverse environmental impacts; or where there is a substantive undesirable effect not contemplated in the approval.

The regulations empower an inspector under Regulation 32 to enter onto any land, premises or other facility related to a project for which a project brief, or an environmental impact assessment has been made under the regulations, to determine how far the predictions made in the project brief, or the environmental impact statement, whichever the case may be, are complied with.

The enforcement of these regulations is through the lead agency and the duties it imposes on any developers including those in the field of agriculture, breach of which there are sanctions such as revocation of certificates among others. The inspectors are empowered to enter onto any land and ensuring compliance with the terms of the certificate of approval is the other way through which the regulations are enforced.

The Prohibition of the Burning of Grass Act Cap 33

The Act prohibits burning of grass in all areas in Uganda without authorisation from a sub county chief, veterinary officer, agricultural officer or forest officer. Bush burning affects soil fertility, a key component to agricultural productivity. The need to monitor and avoid the negative effects of agricultural land use such as soil erosion has formed the basis of the discourse and awareness on soil governance, and has also seen the emergence of science and technology as the link between soil management and governance.

High intensity fires decrease soil productivity and consequently lead to a reduction in soil quality. Fires typically result in the reduction of fuel and organic soil nutrient pool sizes. Fire alters several physical soil properties, such as soil structure, texture, porosity, wettability, infiltration rates and water holding capacity.[26] The extent of fire effects on the soil physical properties depends on fire intensity, fire severity, and fire frequency. Low intensity fires do not cause enough soil heating to produce significant changes to soil physical properties.

Soil porosity can also be reduced by the loss of soil invertebrates that channel in the soil. When fire exposes mineral soils, the impact of raindrops on bare soil can disperse soil aggregates and clog pores, further reducing soil porosity.[27] Fires may also induce the formation of a water repellent soil layer by forcing hydrophobic substances in litter downward through the soil profile. Extensive water-repellent layers can block water infiltration and contribute to runoff and erosion. Formation of water-repellent layers is an important concern in soil health.[28]

2.3   Livestock Management

Livestock management is the practice of efficient and productive care-taking of any agricultural related livestock. Livestock (singular or plural) is any domesticated mammal intentionally reared in an agricultural setting for the purposes of profit or subsistence, whether for food, fiber, dairy, draft, breeding, sport purposes, or other product or labor. As such, livestock includes animals such as cattle, horses, sheep and fur-bearing animals, but does not include farmed birds (turkeys, chickens,

---

[26]Kennard et al. (2008).

[27]Kennard et al. (2008).

[28]Kennard et al. (2008).

pigeons, geese), fish, shellfish, amphibians (frogs), and reptiles. It also does not include animals kept as pets.[29]

There are many different aspects that are involved in proper livestock management. Such things include proper feed rations, correct dosages of medicine, intelligent breeding practices, adequate living conditions and many others that go into properly and efficiently producing livestock. The Livestock Census indicated that the national cattle herd is estimated at 11.4 million, 12.5 million goats, 3.4 million sheep, 3.2 million pigs and 37.4 million poultry birds. The Census also showed that livestock numbers had increased across all animal types: cattle, sheep, goats, poultry and others. But livestock production levels could only meet half of the domestic and regional demand. Growth in the sector has been achieved as a result of a favorable macroeconomic environment, policy and institutional reforms including liberalization of the sub-sector.[30]

The legal and regulatory framework is contained in the following framework. The Animals (prevention of cruelty) Act Cap 39 is important for controlling cruelty against animals. Section 2 provides that any person who cruelly beats, kicks, ill-treats, overrides, overdrives, overloads, tortures or infuriates any animal, or permits any animal to be so used, causes any unnecessary suffering, administers any drug or substance to any animal or kills any animal in an unnecessarily cruel manner, commits an offence of cruelty and is liable on conviction to a fine not exceeding one thousand shillings or to imprisonment for a period not exceeding three months or to both such fine and imprisonment. The Act also puts in place measures to combat the spread of animal diseases. Section 4 provides that an authorized officer may seize any animal suffering from any contagious or infectious disease which is at large in any public place and any court may order for it to be destroyed.

While the Act protects animals against cruelty and disease spread, there is need to amend the act in regard to the punishments since the current ones are not punitive enough. There is also need for sensitisation so that people improve their perception about animals, value animal life and the contribution of livestock to the improvement of livelihoods through agricultural investment.

The Animal Breeding Act 2001 establishes the National Animal Genetic Resources Centre and Data Bank. It provides for the promotion and regulation of animals and fish genetic materials. It also makes provision for the implementation of the national breeding policy in Uganda. Under Section 9 a sample of all genetic materials should be submitted to a national depository for examination and future reference. These include: semen materials, ova eggs and embryos. The 5th Schedule provides for the required performance parameters of animals and fish for collection of semen and sperm.

---

[29]New World Encyclopedia (2008).
[30]UBOS (2018).

Under the Act, the National Animal Genetic Resources Centre and Data bank has the leading role of gradually commercializing the breeding activities; procurement and distribution of semen and equipment and training.

The Animal Diseases Act Cap 38 makes provision with respect to measures to control diseases relating to animals. It requires separation of infected animals from others not so affected and the owner to notify an administrative officer or veterinary officer or inspecting officer.

Section 4 of the Act requires the administrative officer in charge of a district or area, to notify farmers of the outbreak or existence of a disease affecting stock. The officer may also require the animals to undergo a period of quarantine if he or she considers the same to be necessary in order to prevent the spread of diseases. The Act is however difficult to enforce especially in pastoral areas of eastern Uganda where herdsmen move from one place to another.

The Cattle Grazing Act Cap 42 controls and regulates cattle grazing in particular areas in Uganda as the Minister for Agriculture may by statutory instrument declare. Under the Act, cattle means bulls, cows, oxen, calves, sheep and goats. Section 2 of the Act prohibits grazing cattle on any land where the veterinary officer or district administration has prohibited grazing. An order can also be made prescribing the maximum number of cattle that may be grazed on any particular area of land.

Any person who contravenes such orders commits an offence and is liable on conviction to a fine not exceeding one thousand shillings or to imprisonment for a period not exceeding six months or to both such fine and imprisonment. The Act is important for sustainable livestock farming because it sets conditions that should be complied with.

2.4   Crop Production and Management

Crop production management is the organization and coordination of the activities of a branch of agriculture that deals with growing crops for use as food and fiber. Crop production management depends on the availability of arable land and is affected in particular by yields, macroeconomic uncertainty, as well as consumption patterns; it also has a great incidence on agricultural commodity prices.[31] It also depends on the availability of arable land and is affected in particular by yields, macroeconomic uncertainty, as well as consumption patterns.

Crop production management involves programs which increase crop yields while maintaining the environment. Such programs include:

• Conducting research on issues that affect crop breeding, growth and production. The National Agricultural Research Organisation Act 1992 establishes the National Agricultural Research Organisation (NARO) with the goal of enhancing

---

[31]OECD Data (2018).

the contribution of agricultural research to sustainable agricultural productivity, sustained competitiveness, economic growth, food security and poverty eradication in Uganda.

• Reducing soil erosion and improving soil quality through implementing newly developed and time tested methods that improve soil fertility. The most promising and profitable technological option for improving soil productivity is using a combination of organic and inorganic fertilizers, with erosion control measures where necessary. Improved management of existing organic sources, such as methods integrating manure and composting, may significantly increase soil organic matter and reduce nutrient loss.[32]

• Reviving old varieties of crops while developing new varieties. The Uganda National Seed Strategy 2014/15–2019/20, under Strategy 1.1.1 provides for supporting and promoting development and use of new varieties for production and marketing of improved varieties and quality seeds. Strategy 2.1.2 provides that the National Agricultural Research System (NARS) should maintain stocks necessary for the conservation of introduced and local plant genetic material and improved varieties to provide for seed security and mitigate natural disasters.[33]

• Helping farmers to plant crops suitable to local and regional climate and crop management needs. The National Agricultural Advisory Services Act 2001 establishes the National Agricultural Advisory Services (NAADS) as a public agency under the Ministry of Agriculture, Animal Industry and Fisheries with the mandate to increase access by all categories of farmers to agricultural inputs for improved household food and nutrition security and household incomes in line with the Agricultural Sector Strategic Plan and National Development Plan within Uganda's Vision 2040. NAADS is working with Operation Wealth Creation (OWC)[34] to transform the lives of Ugandans through improved agricultural production. They offer general support to farmers through provision of planting materials, stocking materials, farm machinery, implements and value addition equipment.

Crop management is regulated under several policies and laws. The policies and laws seek to promote principles of crop management such as crop breeding, growth and production, reduction of soil erosion and improving soil quality, reviving old varieties of crops while developing new varieties helping farmers to plant crops suitable to local and regional climate and crop management needs.

The Uganda Forestry Policy, 2001 is the main policy for sustainable increase in the economic, social and environmental benefits from forests and trees by the people

---

[32]Olson and Berry (2013).

[33]Government of the Republic of Uganda, Ministry of Agriculture Animal Industry and Fisheries (2015).

[34]OWC is an initiative by His Excellency the President of the Republic of Uganda aimed at engaging the Uganda Peoples Defence Forces (UPDF) in improving household incomes by providing support in the coordination of NAADS activities at community level.

of Uganda, especially the poor and vulnerable.[35] Policy Statement 6 provides for the development of strategies for promotion of tree-growing on farms in all farming systems and innovative mechanisms for the delivery of forestry extension and advisory services. Further, the policy seeks to promote and support farm forestry in order to boost land productivity, increase farm incomes, alleviate pressures on natural forests and improve food security. It recognises important opportunities for tree farming on private land including non-wood products and fruit trees.

The strategies for the implementation of this policy statement include[36]:

- Strengthening the organisation of farmers for better communication and collaboration in the development of farm forestry;
- Building the capacity of farmers to integrate forestry into all farming systems; and
- Disseminating farm forestry advice through decentralized, farmer-driven service delivery mechanisms.

The National Forestry and Tree Planting Act, 2003 enacted to implement the Policy seeks to promote the sustainable use of forest resources, protects forests and forest produce, enhances the productive capacity of forests, promotes tree planting and regulates trade in forest produce as a means of agriculture investment. Section 32 of the Act prohibits forestry activities except in accordance with the management plan or in accordance with a licence. Under the section, one must be authorized to cut, take, work or remove forest produce, clear, use or occupy any land for grazing, livestock farming and planting or cultivation of crops. Forest produce as per Section 3 of the Act includes anything which occurs or grows in a forest and includes among others fruits, seeds, honey and mushrooms. Agricultural investments can therefore not be undertaken in forest reserves without a license.

Sections 21 to 27 of the Act provide for an aspect of private forestry. A private individual can own and manage a private forest. An owner may be able to transfer and mortgage a private forest so long as it is registered and managed under a management plan approved by the authorities. This is an incentive for people who would like to invest in forestry.

The Plant Protection and Health Act, 2015 consolidates the law relating to protection of plants against destructive diseases, pests and weeds, to prevent the introduction and spread of harmful organisms that may adversely affect Uganda's agriculture, the natural environment and livelihood of the people. It is also aimed at ensuring sustainable plant and environmental protection, regulating the export and import of plants and plant products, introduction of new plants and enhances the international reputation of Ugandan agricultural imports and exports.

Under section 3 of the Act it establishes the Phytosanitary and Inspection Service in the Department responsible for Crop Protection which is responsible for the protection of the agricultural resources of Uganda from harmful organisms that exist in the country or could be introduced into the country.

---

[35]Byarugaba (2008).

[36]The Uganda Forestry Policy, 2003.

Section 10 of the Act imposes a duty on every occupier or, in the absence of the occupier, every owner of land to take all such measures to combat specified harmful organisms or to cause them to be combated as well as to prescribe or prohibit the use of specified plant protection substances, plant protection equipment or processes for this purpose.

The Seeds and plant Act, 2006 is intended to promote, regulate and control plant breeding and variety release, multiplication, conditioning, marketing, importing and quality assurance of seeds and other planting materials. Under Section 3 of the Act, a National Seed Board under the Ministry of Agriculture is established to formulate and advise government on the national seed policy, establish a system of implementing seed policies, constantly review the national seed supply, coordinate and monitor the public and private seed sector. Section 8 establishes the National Seed Certification Services as a department responsible for certifying seeds, registration and licensing of seed merchants, accreditation and licensing laboratory seed testing and carrying out field inspection, testing, labeling, sealing and eventual certification. The Act is vital for any seed related investment.

The agricultural legal and regulatory regime aims at prevention of introduction of harmful crops and plants that are likely to destroy the indigenous crops. As such any person investing in crop production should ensure the protection of indigenous crops so that they do not become extinct. In this respect, crops that compromise soil quality are discouraged as the growth of the same would be in contravention of the principles laid down in the National Environment Act in respect to soil quality.

2.5   Regulation of Water for Agriculture

Water is a key factor for production in agriculture investment. Districts and sub-county councils with the assistance of agriculture and livestock extension staff and the water officers provide the necessary back-up and supervision of the water users' activities.

Uganda aspires to transform the agricultural sector from subsistence farming to commercial agriculture.[37] Vision 2040 articulates how investments in agriculture will be guided in order to support efforts towards the national goal of transformation, including the development of major irrigation schemes in the country.

There are several laws and policies that regulate water for agriculture. The major ones include the following:

The Water Act Cap 152 provides for the use, protection and management of water resources. The objectives of the Act include promoting the rational management and use of the waters of Uganda through progressive introduction and application of appropriate standards and techniques for the investigation, use, control, protection, management and administration of water resources.

---

[37]Government of the Republic of Uganda, National Planning Authority (2013).

Section 7 of the Act provides for the general rights to use water for domestic use and to irrigate a subsistence garden by the resident or occupier of land where there is a natural source of water. Section 2 defines "domestic use" to include use for the purpose of human consumption, washing and cooking by persons ordinarily resident on the land where the use occurs, watering not more than thirty livestock units, irrigating a subsistence garden and watering a subsistence fish pond. The Act provides for user permits for the use and supply of water where it involves construction and operation of works as well as discharge of waste into a water body. Section 18 requires any person wishing to construct any works or to take and use water to apply to the director of water development in the prescribed form for a permit to do so. The permit is issued requiring the holder to observe certain standard conditions. These include: not causing or allowing any water to be polluted, preventing damage to the source from which water is taken or to which water is discharged after use, taking precautions to ensure that no activities on the land where water is used result in the accumulation of any substance which may render water less fit for the purpose for which it may be reasonably used, observing conditions prescribed by regulations made under this Act and observing any special condition that may be attached to the permit.[38]

The Water permit system of Uganda provides for six types of permits: Surface water and ground water abstraction, Waste water discharge, Drilling, Hydraulic works construction, Easement Certificates. Using a motorized or a powered pump to abstract (extract) water from any surface water body requires a Surface Water Abstraction permit. Using a motorized or a powered pump to abstract (extract) water from a ground water well/spring requires a Ground Water Abstraction permit. The constructing or operating of any works for impounding, damming, diverting or conveying any surface or underground water or draining, or conveying any surface water or draining any land requires either a surface water abstraction permit or a water abstraction permit or a hydraulic works construction permit.[39] Therefore, large scale/commercial irrigation for agriculture productivity requires the relevant permits. The permit system ensures that the use of water is environmentally friendly and promotes sustainable development and detailed procedures are provided in the water regulations.

The Water resources Regulations 1998 made under the Act define procedures of application and regulation of water abstraction permits. Under regulation 6, the factors to be taken into account by the director when considering an application include the existing and projected availability of water in the area, quality of water, any adverse effect which the facility or allocation or use of water under the permit is likely to have on the existing authorized uses of water, an aquifer or waterway, the drainage regime, the availability of any alternative sources of supply and the need to protect the environment among others.

---

[38]Section 20 of the Water Act Cap 152.
[39]Rawarinda (2016).

The Water (Waste Discharge) Regulations, 1998 also made under the Act provide for the establishment of standards for effluent or waste before discharge into water or on land, prohibition on the discharge of effluent or waste and the requirement for waste discharge permits. Regulation 4 prohibits any discharge of effluent or waste on land or into the aquatic environment contrary to the standards established unless with a permit issued by the Director.

If the discharge of wastes is regulated and such wastes are treated first before their discharge into the water or on land, then soil quality shall be protected, resilience promoted and soil health restored. This consequently shall lead to improvement in the soil quality and fertility leading to an increase in soil productivity.

The recently promulgated Policy, the National Irrigation Policy, 2017 has the vision of transforming agriculture through irrigation development.[40] Its goal is to ensure sustainable availability of water for irrigation and its efficient use for enhanced crop production, productivity and profitability that will contribute to food security and wealth creation.

Priority area 1 under the policy is to enhance investments for irrigation development by public, private and other players. The Government is required to implement the following strategic interventions under this priority area:

- Promoting public investments for irrigation targeting development of irrigation infrastructure, agricultural water harvesting and storage;
- Extending affordable credit facilities and bank loans for development of irrigation projects, especially small-scale irrigation schemes to be executed by local community groups and individual progressive farmers;
- Providing incentives to encourage private sector investment in the irrigation schemes; and
- Identifying potential irrigation investments portfolio with special emphasis on those that could attract private sector investors and/or progressive farmers including smallholders.

The National Water Policy, 1995 provides a frame work for the provision of water of adequate quantity and quality for all agricultural activities and investment needs of the present and future generations. It takes into account economic liberalization, privatization and Decentralization reforms. Its objective is to sustainably manage and develop the water resources in an integrated and sustainable manner so as to secure and provide water of adequate quantity and quality for all social and economic needs of the present and future generations.

Under Section 1.2, the policy is intended to promote development and efficient use of water in Agriculture in order to increase productivity and mitigate effects of adverse climatic variations on rain-fed agriculture, with full participation, ownership and agreement responsibility of users. Chapter 6 of the Policy deals with the water supply policy and strategy as it relates to crops, livestock and fish production.

---

[40]Government of the Republic of Uganda, Ministry of Agriculture, Animal Industry and Fisheries and Ministry of Water and Environment (2017).

The policy objective with regard to water for production is to promote development of water supply for agricultural production in order to modernize agriculture and mitigate effects of climatic variations on rain-fed agriculture through:

- Promoting proper water resource assessment and planning for agricultural production;
- Increasing the capacity of farmers to access and use water for crop, fisheries and livestock production; and
- Promoting appropriate water harvesting technologies for irrigation and livestock development.

The National policy for the Conservation and Management of Wetland Resources, 1995 compliments the goals and objectives of the National Environment Management Policy (NEMP), 1995. The overall goal of the policy is to maintain an optimum diversity of uses and users and consideration for other stakeholders when using the wetland.

The objectives include:

- Establishing the principles by which wetland resources can be optimally used now and in the future;
- Ending practices which reduce wetland productivity;
- Maintaining the biological diversity of natural or semi-natural wetlands;
- Maintaining wetland functions and values; and
- Integrating wetland concerns into the planning and decision making of other sectors.

In relation to agriculture, the relevance of the policy is premised on the need to protect wetland systems, maintenance of biodiversity and the environmental impact assessment requirement for all proposed projects in wetland areas. The major weakness of the policy is that it emphasizes the role of the central government and local government in the management of wetlands and omits that of individuals, private sector and communities.

As a person investing in water for agriculture, it is incumbent on such an investor to ensure that activities such as installation of irrigation machinery do not in any way affect the environment but rather practice best practices that are environmentally friendly. In case any of such activities are likely to negatively impact on the environment, investors are enjoined to see to it that they require the necessary permits under the National Environment Act and the regulations thereunder. Where the required permits as earlier discussed are not obtained, then soil quality shall deteriorate, and as such the government and stake holders and the law put in place shall have failed to properly govern soil use.

2.6   Pesticides Control

Pests are unwanted and destructive insects or any animals that attack food or livestock both during the growing and post-harvest seasons. Pesticides are often used to control pests. Common pests in Uganda include weevils, locusts and caterpillars while diseases include coffee wilt, banana wilt and cassava mosaic.[41] While pesticides play an important role in sustaining food supply, they may also be hazardous to human health and the environment if not used as intended.[42] According to Agro Input Policy Brief (2010), it is estimated that counterfeit and fake agro chemicals account for about 10% to 15% of the national agrochemicals in the market valued at USD 6 million per year. The control and regulation of agro chemicals is vital because the unregulated use of agro chemicals is detrimental for the environment and exposes ecosystems to considerable damage.

There are several policies and laws that regulate application of pesticides in the agricultural sector.

The National policy for disaster preparedness and management, 2010 notes that pest numbers increase due to one or a combination of ecological factors including among others, temperature, monoculture, introduction of new pest species, weak genetic resistance, poor pesticide management, bad weather patterns, and migration. Pests lead to damage of plants and harvested crops, consequently leading to food shortages, famine and economic stress. Risk can be reduced through pest monitoring and using an integrated pest management approach.[43]

Policy Statement 2.1.9 requires all stakeholders to take the following actions:

- Promote research into pest resistant crops;
- Carryout surveillance of crop diseases and monitoring of crop production;
- Ensure spraying of crops; and
- Promote proper post-harvest crop husbandry.

This policy is important because it guides the control of pests and pesticides so as to boost agricultural productivity while at the same time ensuring that the environment and associated eco-systems, human health and socio-economic development are protected through the rational use of pesticides.

The Agricultural Chemicals (Control) Act 2006 controls and regulates the manufacture, storage, distribution and trade, use, importation and exportation of agricultural chemicals and for other related matters.

Section 2 of the Act defines "agricultural chemicals" to include plant protection chemicals, fungicides, insecticides, nematicides, herbicides, matricides, bactericides, rodenticides, molluscides, avicides, fertilizers, growth regulators, wood preservatives, bio-rationals, bio pesticides, bio-fertilizers or any other chemicals used for

---

[41]Government of the Republic of Uganda, Directorate of Relief, Disaster Preparedness and Refugees (2010).

[42]Uganda Pesticides Control Board (2018).

[43]Ludwig et al. (2018).

promoting and protecting the health of plants, plant products and byproducts. Section 5 establishes the Agricultural Chemicals Board that is required to ensure that agricultural chemicals are duly registered and used in a manner consistent with the labeling and in conformity with the regulations made under the Act. The Act therefore plays a big role of controlling agricultural chemicals in promoting agricultural investment.

The Control of Agricultural Chemicals (Registration and Control) Regulations 1989 requires the Agricultural Chemicals Board to register agricultural chemicals, fumigators, commercial applicators and premises. The regulations give restrictions on storage, use, safety, disposal of agricultural chemicals and safeguard of the environment. Agricultural investment requires compliance with the necessary chemical standards set under the Act and the Regulations.

Section 10(2) of the Investment Code Act Cap 92 prohibits foreign investors from carrying on the business of crop production, animal production or acquiring or being granted or leased land for the purpose of crop production or animal production. The section however does not prohibit a foreign investor from providing material or other assistance to Ugandan farmers in crop production and animal production or leasing land for purposes of manufacturing or carrying out the activities set out in the Second and Third Schedules to the Act. Under the second schedule to the Investment Code Act Cap 92, crop processing, processing of forestry produce, fish processing, textile and leather industry, paper production and meat processing are some of the priority areas for investment in Uganda.

In accordance with the National Environment Act Cap 153 any investor who wishes to introduce any agricultural chemical that is likely to have an impact on the environment and soil quality is required like any other developer to undertake an environmental audit and to also carry out an environmental impact assessment to ensure that such chemical does not affect the environment and soil quality.

Harmful chemicals affect the soil quality and health hence the arching need to have laws and policies in place to govern the use of soil in the country.

2.7   Fertilizers Management

Fertilizer management in the agricultural sector is regulated by the National Fertilizer Policy 2016 which synthesizes the fragmented policies of the Government on fertilizer into a single coherent whole. The objective of the Policy is to support the sub-sector in ensuring that the fertilizer industry provides affordable and accessible fertilizers to farmers for increased and sustainable agricultural productivity and farm incomes.[44]

---

[44]Government of the Republic of Uganda, Ministry of Agriculture Animal Industry and Fisheries (2016).

The National Environment Act Cap 153 under the third schedule of the Act requires EIA to be conducted for large scale agricultural projects, use of pesticides, introduction of new crops and use of fertilizers.

2.8 Regulation of Investment in the Agriculture Sector

Investment in agriculture like other sectors is generally regulated under some laws. The Investment code Act Cap 92 regulates both local and foreign investments in Uganda. Section 10 of the Act prohibits foreign investors from carrying on the business of crop production, animal production or acquiring or being granted or leased land for the purpose of crop production or animal production. It restricts foreign investors to provide material or other assistance to Ugandan farmers in crop production and animal production or lease land for purposes of manufacturing or carrying out the activities set out in the Second and Third Schedules to Act. Among these activities are crop processing, fish processing, meat processing, packaging industry, and manufacturing of tools, implements, equipment and machinery.

The Act under Section 21 stipulates that an investor who imports any plant, machinery, equipment, vehicles or construction materials for an investment project shall benefit from the concessional rates of import duty and other taxes as may be specified in the Finance Acts from time to time.

The Act provides for enterprises which qualify for incentives under the code. Under Section 22 an investor in a business enterprise who commences operationafter the coming into force of the Code shall qualify for incentives if he or she satisfies three or more of the following objectives:

- the generation of new earnings or savings of foreign exchange through exports, resource-based import substitution or service activities;
- the utilisation of local materials, supplies and services;
- the creation of employment opportunities in Uganda;
- the introduction of advanced technology or upgrading of indigenous technology;
- the contribution to locally or regionally balanced socioeconomic development; or
- any other objectives that the authority may consider relevant for achieving the objects of the Code.

Where an investor is a citizen of Uganda and the value of his or her investment is at least fifty thousand United States dollars, such enterprise qualifies for incentives.

Under Section 23 an investor who intends to avail himself or herself for incentives under the Act may, if qualified in accordance with section 22, apply to the authority for a certificate of incentives.

A holder of a certificate of incentives shall be entitled to a drawback of duties and sales tax payable on imported inputs used in producing goods for export as provided in any law imposing such duties or taxes.[45] These incentives do not set prerequisites for local and foreign investment but rather encourage and boost investment in Agriculture by both local and foreign investors in accordance with the code.

2.9   Taxation in the Agricultural Sector

The Agricultural sector is subjected to taxation. Under Section 35 of the Income Tax Act Cap 340, a person carrying on the business of horticulture in Uganda to produce income included in the gross income and such person that has incurred expenditure of a capital nature on the acquisition or establishment of a horticultural plant or the construction of a greenhouse shall be allowed a deduction of an amount equal to twenty percent of the amount of expenditure in the year of income in which the expenditure was incurred and in the following four years of income in which the plant or greenhouse is used in the business of horticulture carried on by the person.

The Value added tax (VAT) Act, Cap 349 is a tax on consumption which is imposed on the supply of goods and services (taxable supplies) made by a taxable person other than exempt supplies and imports other than exempt imports.[46] Under the Act the standard rate of VAT is 18%. The Act provides for the following tax incentives essential for agricultural investment.

Under Section 19 of the Value Added Tax Act Cap 349, provision is made for exempt supplies which are listed in the second schedule to the Act. These exempt supplies include the supply of livestock, unprocessed foodstuffs and unprocessed agricultural products except wheat grain. They further include the supply of unimproved land, veterinary services and the supply of machinery, tools and implements suitable for use only in agriculture.

Such tools include knapsack sprayers, ox ploughs, drinkers and feeders for chicken, agricultural tractors (including walking tractors), disk harrows, cultivators, ploughs, weeders, seeders, planters, subsoilers, seed drills, threshers, bale wrappers, milking machinery, milk coolers, maize mills, wheat flour mills, homogenizers, dairy machinery, grain cleaners and sorters, feed grinders hatcheries and implements used for artificial insemination in animals. Section 24 of the Value Added Tax Act Cap 349 makes provisions for zero rated supplies listed in the third schedule. These include the supply of seeds, fertilizers, pesticides and hoes.

VAT exempt supplies under the second schedule to the Act include:

- Supply of livestock, unprocessed foodstuffs and unprocessed agriculture products except wheat grain;
- Supply of unimproved land; and

---

[45]Section 25 of the Investment Code Act.
[46]Uganda Revenue Authority (2013).

- Supply of machinery, tools and implements for use only in agriculture, i.e. planters, harvesters, seeders, weeders, hoes, knapsack sprayers, irrigators, cultivators, ploughs, fertilizers, dairy machinery, agriculture tractors etc.

Persons dealing only in exempt supplies are not expected to register for VAT. Zero rated supplies under the third schedule to the Act include the supply of seeds, fertilizers, pesticides, and hoes. Persons dealing in zero rated supplies are expected to register for VAT in case they meet the registration requirements. Agricultural Investment opportunities therefore exist in such listed exempt and zero rated items under the VAT Act.

3 Conclusion

Agriculture is the predominant economic activity in Uganda and the government's vision is to transform the agricultural sector from subsistence farming to commercial agriculture. Regulation of the activities in the agricultural sector is very critical because there is need to create standards to regulate land for agricultural purposes, soil governance, livestock management, crop production and management, regulation of water for agriculture, pesticides control, fertilizer management, regulation of investment in the agricultural sector and taxation. Whereas the policy and legal framework has been developed and provides opportunities in regulation of the critical legal issues, there are still challenges of implementing the policies and laws due to lack of knowledge and awareness, gaps in the policies and laws especially incentives of small scale farmers and protection of their health and safety. Therefore, it is important to review the policy and legal framework to address the above critical issues in the agriculture sector.

References

Black HC (1990) Black's law dictionary: definitions of the terms and phrases of American and English jurisprudence, ancient and modern, 6th edn. Minn. West Publishing Co., St. Paul

Byarugaba SR (2008) Forest policy, legal and institutional framework information report for the Republic of Uganda. Available at: http://www.fao.org/forestry/14186-05f8c554e1b2f08f5ebb36f1bdfa02940.pdf

Byrnes RM (1992) Library of Congress. Federal Research Division, Thomas Leiper Kane Collection (1992): Uganda: a country study. Washington, D.C. Available at: https://www.loc.gov/item/92000513/

Conserve Energy Future (2018) Causes effects, solutions of soil erosion. Available at: https://www.conserve-energy-future.com/causes-effects-solutions-soil-degradation.php

Encyclopedia of the Nations (2018) Uganda – Agriculture. Available at: http://www.nationsencyclopedia.com/economies/Africa/Uganda-AGRICULTURE.html

Government of the Republic of Uganda (2010) The National policy for disaster preparedness and management. Available at: https://www.ifrc.org/docs/IDRL/Disaster%20Policy%20for%20Uganda.pdf

Government of the Republic of Uganda (2015) The Uganda National Seed Strategy 2014/15–2019/
    20. Available at: http://extwprlegs1.fao.org/docs/pdf/uga175068.pdf
Government of the Republic of Uganda (2017) National Irrigation Policy. Available at: https://
    www.mwe.go.ug/sites/default/files/library/Uganda%20National%20Irrigation%20Policy.pdf
Government of the Republic of Uganda, Ministry of Agricultural Animal Industry and Fisheries
    (2016) National Fertiliser Policy. Available at: http://extwprlegs1.fao.org/docs/pdf/uga172925.pdf
Government of the Republic of Uganda, Ministry of Agriculture, Animal Industry and Fisheries
    (2013) National Agriculture Policy. Available at: http://agriculture.go.ug/wp-content/uploads/
    2019/04/National-Agriculture-Policy-1.pdf
Government of the Republic of Uganda, National Planning Authority (2013) Uganda Vision 2040.
    Available at: http://npa.go.ug/wp-content/themes/npatheme/documents/vision2040.pdf
Kabanda N (2013) Uganda's National Land Policy. Background, Highlights and Next Steps. Focus
    on Land in Africa. Available at: http://www.focusonland.com/fola/en/resources/ugandas-
    national-land-policy-background-key-highlights-and-next-steps/
Kalyango R (2012) 10 coffee factories closed over poor handling methods. The New Vision June
    5th 2012. Available at: https://www.newvision.co.ug/new_vision/news/1302251/coffee-facto
    ries-closed-poor-handling-methods
Kennard D, DiCosty RJ, Callaham MA Jr (2008) "Fire effects on soil", Forest Encyclopedia
    Network. Available at: http://www.forestencyclopedia.net/p/p622
Lal R, Hall GF, Miller FP (1989) Soil degradation: I. Basic processes. Land Degrad Dev 1:51–69.
    https://doi.org/10.1002/ldr.3400010106
Ludwig M, Wilmes P, Schrader S (2018) Measuring soil sustainability via soil resilience. Sci Total
    Environ 626(18):1484–1493. https://doi.org/10.1016/j.scitotenv.2017.10.043
Nations Encyclopedia (2011) Uganda. Available at: https://www.nationsencyclopedia.com/Africa/
    Uganda.html. Accessed 16 Dec 2019
Negm AM, Abdel-Fattah S (2017) Grand Ethiopian Renaissance Dam Versus Aswan High Dam.
    The handbook of environmental chemistry. Springer International Publishing
New World Encyclopedia (2008) Livestock. Available at: http://www.newworldencyclopedia.org/
    entry/Livestock
Olson J, Berry L (2013) Land Degradation in Uganda: its extent and impact. Available at: https://
    rmportal.net/library/content/frame/land-degradation-case-studies-05-uganda/at_download/file
Organization of Economic Cooperation and Development (2018) OECD Data: Agricultural Output-
    Crop Production. Available at: https://data.oecd.org/agroutput/crop-production.htm
Pesticide Control Board of the Republic of Uganda (2018) Rational Pesticide Management for
    Sustainable Food Systems
Rawarinda ME (2016) The Water Permits System (WPS) in Uganda
Trading Economies (2019) Uganda GDP from Agriculture. Available at: https://tradingeconomics.
    com/uganda/gdp-from-agriculture
UBOS (2018) Statistical Abstract, UBOS, Kampala, Uganda
Uganda Revenue Authority (2013) What is Value Added Tax? Available at: https://www.ura.go.ug/
    Resources/webuploads/INLB/Value%20Added%20Tax.compressed.pdf

Land Conservation in the Albertine Graben Region of Uganda: A Critical Analysis of the Legal Regimes

Wahab Kassim

1 Introduction

In 2006 the Ugandan government declared that the country was commercially viable for production of oil and gas[1] and that the Albertine Graben was identified as the most prospective amongst all the other basins.

Prior to oil discovery, the Albertine region was well known as a protected area for wildlife. It was well known as the area for International Bird, Ramsar and World Heritage Sites.[2]

However, the oil industry presents several environmental challenges that Uganda has to deal with, Uganda's economy is known for acquiring revenues from tourism and farming which both are currently at risk due to the dangerous nature of the oil industry. Infrastructures such as refineries, pipelines, central processing units and roads are requirements for the development of the industry, unfortunately these infrastructures have contributed greatly to the degradation of the environment and the land in the Graben.

Scenarios such as deforestation due to the development of oil infrastructure cause land and environmental degradation problems plus loss of biodiversity since land is left bare. This exposes it to erosion especially given the increased water runoff and wind speed.[3] In addition, there is pollution of land due to oil spills, discharge of effluents and poor waste disposal and management. These impacts on land are known for reducing land organic matter and macro nutrients such as nitrogen and

---

[1] Netherlands Commission for Environmental Assessment (2012).

[2] Netherlands Commission for Environmental Assessment (2012), p. 3.

[3] Kakura and Ssekyana (2009), p. 65.

W. Kassim (✉)
Kampala International University, School of Law, Kampala, Uganda

© Springer Nature Switzerland AG 2020
H. Yahyah et al. (eds.), *Legal Instruments for Sustainable Soil Management in Africa*, International Yearbook of Soil Law and Policy,
https://doi.org/10.1007/978-3-030-36004-7_5

phosphorus and consequently introducing metals, toxins, acidification and depletion of bases.[4] Ling et al. have enumerated that food chains are affected due to heavy metal contamination of the land which poses a risk to humans.[5]

On the basis of the impacts the oil activities and infrastructure have on land, this chapter makes a critical analysis of the several laws available to safe guard the Albertine environment against the deleterious impacts.

The 1995 Ugandan Constitution provides inter-alia for the rights to property and compensation in case of acquisition,[6] every Ugandan has a right to a clean and healthy environment,[7] citizen participation in development and their own governance.[8]

These constitutional provisions are backed up by other laws and policies including the Land Act,[9] Forestry Act, the Petroleum (Exploration, Development and Production) (PEDP) Act of 2013 (upstream) and midstream the Petroleum (Refining, Conversion, Transmission and Midstream Storage) (PRCTMS) Act 2013.

Policies include the National Land Policy (2013), Oil and gas Policy of 2008 and other important documents such as the National Development Plan that specifically recognizes the need to strengthen the policy, legal and regulatory framework for the oil and gas sector.

Existing environmental legal framework in Uganda is potentially in conflict, relatively old and outdated, there is urgency to amend most of the laws and policies so that they can address the new challenges such as protection of wildlife, forest reserves and human settlements from impacts of the oil industry for example carbon emissions, gas flaring and waste management.

Notably Uganda has not joined the Extractives Industry Transparency Initiative (EITI) but is signatory to various international agreements and conventions, which are meant to protect and conserve the environment, society, biological diversity, natural resources, wetlands, culture and human rights among others.

This chapter contains four sections. Section 2 of this chapter makes a comprehensive description of the Albertine Graben environment and the historical context of oil and gas activities, Sect. 3 states the impacts of oil and gas activities on the Albertine land Sect. 4 scrutinizes the efficacy of the relevant legal regimes and the Ugandan constitution in protecting and conservation of land and Sect. 5 of the chapter avails recommendations attained from the identified practices for enforcement of land preservation and sustainable usage.

---

[4]Nkonya et al. (2002), p. 10.

[5]Ling et al. (2007).

[6]Article 26(2)(b)(i) of the 1995 Constitution.

[7]Article 39 of the 1995 Constitution.

[8]Principle II of the Democratic principles under the National Objectives and Directive principles of state policy in the 1995 Constitution of Uganda.

[9]Land Act 1998 (Cap 227).

2 Background

2.1 Historical and Geological Status of the Albertine Graben

Uganda has five sedimentary basins including the Albertine Graben, Hoima Basin, Lake Kyoga Basin, Lake Wamala Basin and Kadam-Moroto Basin.[10] The focus of this chapter is the Graben that is estimated to be holding over 4.5 billion barrels of crude oil thus making it the fourth largest oil reserve in Sub Saharan Africa.[11] It is situated in the northern area part of the western part of the East African Rift Valley System, 500 km long, averaging 45 km wide and 23,000 square km.[12] The Graben strides between Uganda and Sudan in the north to Lake Edward in the south, (including the Democratic Republic of Congo).[13]

According to the National Environment Management Authority (NEMA) sensitivity Atlas

The Graben is a Cenozoic rift basin formed and developed on the Precambrian orogenic belts of the African Craton. Rifting was initiated during the late Oligoceneor Early Miocene.

The Albertine Graben was initiated by either reactivating pre-Cambrian lineaments or creating new normal faults by an extensional regime during the Cretaceous prior to initiation of the East African Rift System. This was followed by a compressive regime and this regime could have persisted into late Oligocene or even Early Miocene as evident from seismic data. The earliest dated sediments from Semliki basin indicate an age of Early Miocene. The Graben trends in a North East to South West direction through most of its length. Each of the rift basins in the Albertine Graben is bounded by steep border faults and broad uplifted flanks (escarpments) that are predominantly pre-cambrian basement composed of metamorphic rocks such as gneisses, quartzite, schist and varying amounts of mafic intrusions.[14]

The Graben is known for its pristine natural environment, rich wildlife and community wildlife reserves coupled with eco-tourism in Murchison Falls National Park grossing a lot of revenue annually for Uganda. It is reported that in 2009 the national park was visited by approximately 40,000 visitors.[15]

The Albertine area is currently recognized globally as a biodiversity hotspot, containing over 50% of birds, 39% of mammals, 19% of amphibians, and 14% of reptiles and plants found in mainland Africa.[16] Unfortunately Murchison Falls National Park lies in Exploration Area 1, in which 6 wells (Mpyo 1, 2 & 3 in the

---

[10]Global Rights Alert (2017).

[11]International Monetary Fund (2010).

[12]NEMA (2009), p. 3.

[13]NEMA (2009), p. 3.

[14]NEMA (2009), p. 3.

[15]Johnson et al. (2014).

[16]Plumptre et al. (2009).

south section of the Park and Rii, Jobi and Jobi-East 1 in the north side of the Park) have been drilled.

Several lakes,[17] perennial and seasonal rivers such as Hohwa River, Wambabya River,[18] Sebugoro, Kabyosi, Warwire and Nyamasoga feed and drain in Lake Albert also ranked as Africa's seventh largest lake consequently all these water resources are close to oil wells.[19]

2.2   Land in the Graben

The Graben has fertile land that is used for several purposes including national parks, wildlife reserves and forest reserves), agriculture (crops and livestock), plant species and human settlements.[20]

Albertine Land harbors 10 of the 39 wildlife protected areas Uganda boasts of, amongst of these national parks include Murchison Falls, Queen Elizabeth, the Rwenzori Mountains, Kibale, Semliki, Bwindi and Mgahinga[21] and wildlife reserves (Ajai and East Madi,[22] Bugungu and Karuma,[23] Tooro-Semliki, Kabwoya and Kyambura,[24] and Kigezi Wildlife Reserve).[25]

Several forest reserves are also situated in the Albertine such as Bugoma and Budongo Forest Reserves.[26] The Albertine is said to hold at least 5793 different plant varieties recorded in the region, and 551 of these species are endemic to it.

Apart for the reserves, Albertine has settlements and populations that seek sustenance from the land through agriculture, livestock and settlement. Land tenure in the Graben is predominantly customary with a few areas of Mailo land and freehold tenures.

It's also claimed that customary land tenure is likely to be stable and secure since land is passed from one generation to another in a smooth and stable manner.[27] However it's been criticized for its weakness in conservation and management of common resources, conversion of land from woodland to crop production[28] and high

---

[17]There are three main lakes; Lake Albert, Lake Edward, and Lake George.

[18]The perennial rivers (Hohwa and Wambabya) flow continuously with peak flow during the rainy season.

[19]NEMA (2009), p. 9.

[20]NEMA (2009), p. 12.

[21]NEMA (2009), p. 13.

[22]Located in the extreme north-east of this region.

[23]Situated in Buliisa and Masindi Districts respectively which are located in the immediate south.

[24]Located mid-way of the region.

[25]Situated in the extreme south-west.

[26]NEMA (2009), p. 13.

[27]Nkonya et al. (2002).

[28]Nkonya et al. (2002), p. 5.

population growth and influx has been witnessed.[29] The population growth and influx has precipitated land grabbing plus wealthy individuals processing communal land titles into private freehold titles.

In effect, the Graben is rich in environmental and wildlife resources and this demands strict implementation of conservations laws so that oil exploration and development activities do not harm the land and the rich ecology. Oil reserves are finite, it is pertinent that oil resource extraction does not compromise the fertile land since livelihood of the communities living in the Graben were dependent upon the biodiversity.[30]

Ever since government embarked on several physical infrastructures to promote the oil industry, several developments are ongoing including construction of roads, an international airport, a 60,000 barrels-per-day Greenfield refinery.[31]

With these developments, the scenic environment is at threat since Uganda's oil is waxy, infrastructure requirements are larger.[32] Extra power plants are required to conduct heating, storage, and transport of oil plus the shallow depths of oil wells and weak natural flow pressures will require significant water injection for oil extraction.[33] Most water injected in the wells is extracted from Lake Albert, water demand is at the peak during the construction phase of the key projects estimated at 75,000 ml per day.[34] Consequently it is not yet clear where the released waste water which contains oil deposits will be stored and recycled.

To exacerbate issues, leaders in the Albertine area have also decried the increasing cases of land grabbing as well as irregularities in the process of compensating people whose land and properties are affected by oil and gas infrastructural projects.[35]

2.3   Oil Exploration in the Graben

Statements on Oil seepages in Uganda were initially availed by locals and later E.J. Waylands a government geologist further documented substantial amounts of hydrocarbons in the Albertine Graben.[36]

The first wells, Waki-B1 were drilled by the Anglo European Investment Company of South Africa in 1938.[37] With the emergency of World War II, the oil and gas

---

[29]Golombok and Jones (2015), p. 3.

[30]Oil in Uganda (March 6th, 2018).

[31]The Independent (July 18th, 2017).

[32]Oxford Institute for Energy Studies (2015), p. 36.

[33]Oxford Institute for Energy Studies (2015), p. 36.

[34]MEMD (2017), p. 5.

[35]Oil in Uganda (March 6th, 2018).

[36]Johnson et al. (2014), p. 26.

[37]SNF (2011).

industry stagnated coupled with the colonial tendencies and policies[38] in Uganda that never favored exploration at the time.

Later on in the 1940s and 1950s further exploration was conducted, several shallow wells were drilled mainly for stratographic purposes.[39] Around 1980s intensive exploration work was conducted plus aeromagnetic data in 1983 confirmed the existence of viable sedimentary basins in the Albertine Graben, Lake Kyoga basin, Lake Wamala basin and Moroto Kadam basin. With these expectations government enacted the Petroleum (Exploration and Production) Act in 1985 to streamline exploration and production of petroleum and related matters.[40]

The Act laid a foundation towards licensing several international companies plus the subsequent seismic surveys and drilling. When the Petroleum Exploration and Production Department (PEPD) replaced the Petroleum Unit, the PEPD conducted aeromagnetic surveys[41] and in 1993 comprehensive petroleum (exploration and Productions Regulations) were passed. Since then international companies that have embarked on exploration include Fina (Production Sharing Agreement signed with government in 1991), Heritage oil & Gas Ltd in 1997 for exploration Area 3, Australia's Hardman Resources in 2001 for exploration Area 2, Dominion Uganda Ltd, Energy Africa and Neptune Petroleum.[42]

Currently there are three major companies conducting extensive exploration and have Production licences in the Albertine and include Tullow Uganda Operations pty LTD, CNOOC (U) LTD, Total E&P Uganda B.V and in a PSA signed on 14th September 2017 an exploration Licence was issued to Armour Energy Ltd for the 344 square km kanywataba field.[43]

3 Oil Development and Associated Impacts on Land in the Graben

Developments such as roads, airports, pipelines and refinery infrastructures degrade land and fragment habitats in the Albertine Graben. All these developments have different impacts on the fertile land from chemical materials such as lead, chlorinated hydrocarbons, heavy metals, zinc, arsenic and benzene.

---

[38]Bainomugisha et al. (2006).
[39]Gordon et al. (2011).
[40]Kasimbazi (2012).
[41]Kasimbazi (2012).
[42]Johnson et al. (2014), p. 28.
[43]MEMD (2017), p. 11.

3.1 Oil Infrastructure in the Graben

3.1.1 Roads, Oil Rigs, and Other Infrastructures

In a report[44] to parliament the Minister in charge of Energy and Mineral Development stated that about five roads had been identified and feasibility studies plus Environmental and Social impact Assessments (ESIA) had been concluded to support the timely production of oil and gas in the country.[45]

However laying seismic lines and construction of roads damages the mangroves, which can take more than thirty years to recover.[46] The roads also disrupt the hydrology of Albertine mangroves through interrupting the proper flow of water and destroy fish breeding grounds.[47]

Stevens also states that the;

> Dredging processes remove sediment, land creek banks, and vegetation and deposits it alongside the new channel. Accumulated toxins are removed from the remaining land in the channel and the dredged material. The toxins plus loosened sediment flow into and disrupt the aquatic system.[48]

Infrastructural developments in the Albertine have caused several challenges to the land in the Graben and a 2008 NEMA report states that oil activities such as road construction, drilling and movement of heavy machinery interfered with the Eco system, human settlements and wildlife movements, feeding and breeding although Environmental impact assessments had been taken and mitigation measures proposed.[49]

3.1.2 Refinery, Pipeline and Central Processing Facility (CPF)

Refineries are facilities where semi-finished hydro-carbons are refined and turned into high grade products using processes of electrolysis or distillation whereas a central processing facility (CPF) is an industrial facility that separates oil from water and gas and usually also includes provision for water treatment and a power plant. This facility typically covers an area of approximately 2–5 km^2.

Upon refining several refined products are attained including liquefied petroleum gas, gasoline, kerosene, aviation fuel, diesel fuel, fuel oils, lubricating oils, and feedstocks for the petrochemical industry.

---

[44]MEMD (2017), p. 2.

[45]Funding amounting to USD 531.00 million had been approved and to be availed over two financial years.

[46]Stevens (2014).

[47]Stevens (2014).

[48]Stevens (2014).

[49]Kasimbazi (2012).

In the Albertine land has been acquired for the oil refinery, pipelines and CPF plus the Land Acquisition and resettlement Framework (LARF) has been concluded. Feeder pipelines will be set up to move oil from Buliisa district to Kabaale Parish in Hoima district where government plans to set up a holding terminal for crude oil and an oil refinery.[50] Greenfield oil refinery is a 60,000 barrels per day project with investments from other East African nations and other interested companies.[51]

An Inter-Governmental agreement was signed with Tanzania to construct a, 4455 km East African Pipeline of which 288 km is in Uganda to export crude oil from Hoima in Uganda and other inland fields to Tanga Tanzania.[52]

The pipeline will traverse districts of Hoima, Kakumiro, Kyankwanzi, Mubende, Gomba, Ssembabule, Lwengo and Rakai. Environmental and Social Impact Assessments together with Geotechnical and Geophysical Studies will be undertaken in both countries.[53]

However, there have been reports of irregular evictions of settlers. For example, Oil in Uganda, an organization involved and on ground reporting on all activities in the oil industry has documented some residents complaining about intrusion on people's lands without compensation. Residents leaving in Kasenyi village, Ngwedo sub-county (Buliisa district) state that machines conducting studies for the CPF are already working on people's lands even before receiving compensation for their land and other properties.[54]

Pipelines in the Albertine present a potential challenge as the case is in respect to reports from the Niger delta of Nigeria indicate that over 421 spills have occurred due to equipment failures from Shell Petroleum Development Company (SPDC) pipelines linking the Delta, Bayelsa and River states in Nigeria.[55]

3.1.3   Industrial Park and Airport

An Industrial park is a tract of land developed and subdivided into plots according to a comprehensive plan with or without built-up factories, sometimes with common facilities for the use of a group of industries.

Land amounting 29.57 square km is under development in Hoima and Kabale to facilitate an industrial park to host a refinery an international airport, petrochemical

---

[50]Oil in Uganda (March 6th, 2018).

[51]MEMD (2017), p. 7.

[52]Signed in Kampala on 26th May 2017 by Uganda's Minister of Energy and Mineral Development Eng. Irene Muloni, and Tanzania's Minister for Constitutional and Legal Affairs Hon. Prof. Palamagamba John A.M Kabudi.

[53]MEMD (2017).

[54]Oil in Uganda (6th March 2018).

[55]MOEN (2008), p. 10.

industries and energy based industries, among others.[56] A master plan and detailed engineering designs for phase 1 of the kabale international airport were completed, in a cabinet meeting a loan amounting to UGX 1.3 trillion for the airport was approved and the first phase is expected by the end of the 3rd quarter of 2019.

Concurrently the Resettlement Action Plan (RAP) for the industrial area is on-going but there are a number of challenges affecting its smooth completion such as valuation of community land, speculators and valuation rates.

3.2 Impacts of Oil Infrastructure and Activities on Land in the Graben

On a negative note, due to developments in the Albertine, fertile land is degraded, consequently making the achievement of agricultural targets, food security threatened, and disruption of ecosystem functions, which in the long run reduces ecosystem performance, resilience, and stability.[57]

Land fertility depletion is a widespread problem, with evidence of decline in fertility cited by farmers throughout Uganda.[58]

With oil pollution there's reduction in land organic matter and macro nutrients such as nitrogen and phosphorus, but more toxins, acidification and depletion of bases.[59]

On top of degrading the fertile land, the pollutants are associated to deadly health hazards including cardiovascular disease, throat inflammation, chest pain and congestion.

Heavily contaminated land remains unusable for agriculture for several years.[60]

The need for the different infrastructure in the Albertine has created a conflict between welfare and environmental objectives, these developments are associated with reduced area of forest cover.

Oil and gas development harm land as a result of oil spills and waste dumping as well as other, less well-known effects of exploration and production.

---

[56]The Masterplan for the Industrial Park has been developed taking into considerations the expected industries to be established in future, shared utilities and common services, management and financing structure of the park. The Industrial Park will be managed by the Uganda National Oil Company (UNOC).

[57]Mubiru et al. (2017).

[58]Nkonya et al. (2002), p. 10.

[59]Nkonya et al. (2002), p. 10.

[60]Stevens (2014).

3.2.1 Oil Spills

Oil spills are prone to the oil industry operations and can happen at any time. They can occur as a result of machine blow-outs, rigs and pipeline leaks, tanker spills,[61] sabotage,[62] theft of oil, human error, corrosion, engineering error and poor maintenance of infrastructure,[63] natural disasters, causes of third parties and erosion.

The Centre for Energy at the University of Dundee reports that oil tankers and pipelines are annually responsible for the transporting of some 1800 million tons of crude oil around the world and in majority of cases the transport is done in a relatively safe manner.[64]

Around August and December 2008, two major oil spills disrupted the lives of the 69,000 or so people living in Bodo, a town in Ogoni land in the Niger Delta.[65] Both spills continued for weeks before they were stopped. Estimates suggest that the volume of oil spilled was as large as the Exxon Valdez Accident in Alaska's Prince William Sound. While the big disasters get a lot of attention such as the small scale pollution occurs all the time all over the country, without being noticed.

Amnesty international reports that oil spills in the Niger Delta have caused massive pollution of land of thus destroying crops and damaging the long term productivity of the land.[66]

Oil spills introduce heavy metals such as Cu, Ni and Hg into the land, these metals impair and disrupt the biota of land due to their toxicity.[67] Physical-chemical properties of the land is affected, such as temperature, structure, nutrient status and pH.[68]

Several pollution instances have resulted to litigation to seek remedies for example In Common Wealth of Massachusetts V Andrus[69] the first circuit court of appeals considered a preliminary injunction, issued by the district court, which enjoined the Secretary of interior from proceeding with the proposed sale of oil leaseholds in the OCS off the coast of New England. An injunction had been issued on the basis that safeguards against oil spills were lacking.

Besides destruction of land, oil spills may endanger shorelines, aquatic life, wildlife, and the stability of the entire ecosystem.[70]

---

[61]Laitos and Toumain (1992), p. 448.

[62]A number of spills are attributable to sabotage which is defined as a willful attempt to disrupt or interrupt the production or distribution of oil by third parties.

[63]Oil experts assert that many pipelines and installations in the Niger Delta have not been adequately maintained, and this is a contributory factor in corrosion and leaks.

[64]CEPMLP (2008), p. 44.

[65]Eyinla and Ukpo (2006).

[66]Amnesty International (2009), p. 17.

[67]Ekundayo and Obuekwe (2000).

[68]Ekundayo and Obuekwe (2000).

[69]United States Court of Appeal for the First Circuit- 594 F.2d 872 (1979), 283, 448.

[70]Bradbrook et al. (2003), p. 448.

In Common Wealth of Puerto Rico V SS Zoe Colocotroni,[71] the federal Court of Appeals upheld in part a district court's award of damages for the cleanup costs of an oil spill against a tanker.

The effects of an oil spill are immense thus the need to set stringent measures to avert it is prudent for any state. In the US the Exxon Valdez spill led to the passing of the Oil Pollution Act,[72] the Santa Barbara Spill[73] was to the National Environment Environmental Policy Act (NEPA)[74] and the Coastal Zone Management Act.[75]

3.2.2 Waste Dumping

The Albertine Environment necessitates for proper waste management to avoid polluting the land, wildlife and the general ecology of the area. During oil activities, produced water plus various well treatment chemicals added during production and extraction are trapped underground along with crude that eventually rises to the surface and later separated as part of the oil extraction process. The produced water mixed with oil and grease is then discharged overboard or injected back for secondary oil recovery.

The processes produce solid and liquid wastes that contain materials such as grease, phenolic compounds, cyanide, sulphide, suspended solids, chromium, and biological oxygen demanding organic matter.[76]

In 2012 NEMA issued Operational Waste Management Guidelines for oil and gas operations. They seek to decide on the most appropriate disposal route, both environmental and economic costs and benefits need to be considered. The decision should be reached taking into account all the costs and impacts associated with waste disposal, including those associated with the movement of waste.[77] Wherever possible the Proximity Principle should be considered, it recognizes that transporting waste has environmental, social and economic costs so as a general rule; waste should be dealt with as near to the place of production as possible.

---

[71]United States Court of Appeals for the First Circuit- 628 F.2d 652 (1981).

[72]Often called the OPA 90 passed by senate without a single dissenting vote, even though attempts to strengthen oil spill laws had foundered for years until then. Perhaps the extent of the Exxon Valdez spill juxtaposed against the west and east coasts, explains the unanimity.

[73]Occurred January 28th, 1969 as Union Oil Drilled the 5th well on it Platform A in the channel, the well blew out. Eleven days passed before the well was plugged and some 24,000 to 71,000 barrels of oil spilled into the channel and onto nearby beaches.

[74]Enacted 1969.

[75]Coastal Zone Management Act 1973.

[76]Stevens (2014).

[77]Operational Waste Management Guidelines for Oil and Gas Operations, p. 3.

4 Analysis of the Legal Regime

This part makes a critical analysis of the available laws including the National Environment Act (NEA),[78] Land Act, and Regulations, Forestry Act Petroleum Supply Act 2003, Petroleum (Exploration, Development and Production) Act 2013, (PEDP) and the Petroleum (Refining, Conversion, Transmission and Midstream Storage) Act, 2013, (PRCTMS). Policies include; Uganda National Environment Management policy (NEMP) (1994), The National Policy for the Conservation and Management of Wetland Resources (1995),National Water Policy (1999), National Oil and Gas Policy,[79] and Energy Policy of Uganda.

4.1 Laws

4.1.1 1995 Constitution

The 1995 Constitution states inter alia every Ugandan has a right to a clean and healthy environment.[80]

Section 44(1) of the Land Act reiterates Article 237(2)(b) of the Constitution that provides that Government or a local government shall hold in trust for the people and protect natural lakes, rivers, wetlands, forest reserves, game reserves, national parks and any land to be reserved for ecological and touristic purposes for the common good of all citizens.

The constitution further explicitly enumerates that every person has a right to own property either individually or in association with others.[81]

Whereas government is authorized to acquire land for public interest,[82] the same constitution states that prompt payment and adequate compensation prior to the taking of possession or acquisition of property.[83] Therefore with this provision, a person's right to land is protected from derogation of any nature more so dispossession without adequate compensation.

Consequently, reports have emanated where government officials have attempted to disposes people of their land without adequate compensation.

One Youth leader stated that "Whereas government can acquire our land for projects in public interest, we are entitled to prior, adequate and timely compensation

---

[78]The National Environmental Act, Cap 153 of 1995.

[79]National Oil and Gas Policy for Uganda, NOGP (2008).

[80]Article 39 of the 1995 Constitution.

[81]Article 26 (1) of the 1995 Constitution.

[82]Article 26(2)(a).

[83]Article 26(2)(b)(i).

as enshrined in the constitution. If oil companies do not respect our rights, we shall seek legal redress" said Gilbert Tibasiima the district youth counselor.[84]

Around January 2018 the lands Minister disclosed to residents that compensation for an acre of land in Kasinyi village one of the Albertine areas will be at 3.5 million[85] (approximately USD 821) whilst residents are claiming a compensation of 10 to 60 million per acre[86] (USD2740-16,429) which is at least fair compensation.

Monetary compensation in regards to issues of land are secondary but the primary aspect is that people attach their whole lives to land and once its lost then ones livelihood is at stake. Aldo L. states that "it is inconceivable to me that an ethical relation to land can exist without love, respect, and admiration for land, and a high regard for its value. By value, I of course mean something far broader than mere economic value; I mean value in the philosophical sense."[87]

Therefore disregard of the constitutional provisions is not merely breach but also dislodgement of an important asset for human sustenance thus its paramount that compensation be adequate for anything is right when it tends to preserve the integrity, stability and beauty of the biotic community. It is wrong when it tends otherwise.[88]

4.1.2 1995 National Environment and Management (NEA) Act[89]

The National Environment Act provides for sustainable management of the environment and also establishes an environmental watchdog the National Environment Management Authority (NEMA).[90] NEMA as the principal environmental agency is responsible among others for the management of the environment in Uganda through coordinating, monitoring and supervising of all activities in the field of environment.[91]

NEMA is also mandated to manage environmental information and implementing standards for environment information. Both the Policy and Act are further reinforced by Articles 39.[92]

The roles of NEMA have also been declared in several rulings of court, in *Amooti Godfrey Nyakana V NEMA*,[93] court stated that S.6 of the National environment Act sets out the functions of NEMA inter alia;

---

[84]Oil in Uganda (March 6th, 2018).

[85]Oil in Uganda (March 6th, 2018).

[86]Oil in Uganda (March 6th, 2018).

[87]Leopold (1989), p. 223.

[88]Leopold (1989), p. 223.

[89]1995 National Environment Act, Chapter 153.

[90]Section 4 NEA.

[91]Section 6 NEA.

[92]Article 39 of the 1995 constitution provides for the right to a clean and healthy environment.

[93]Constitutional Petition No. 03/05.

- To coordinate the implementation of government policy and the decisions of the policy committee,
- To ensure observance of proper safeguards in the planning and execution of all development projects, including those already in existence, that have or likely to have significant impact on the environment determined in accordance with the NEA
- To propose environmental policies and strategies to the policy committee,
- To litigate legislative proposals, standards and guidelines on the environment in accordance with the Act,
- To review and approve environmental impact assessments and environmental impact statements submitted in accordance with the NEA or any other law.[94]

NEMA in consultation with the lead agency, is required to establish criteria and procedures for the measurement and determination of land quality; minimum standard, for the management of the quality of the land for their purpose.[95]

NEMA has to issue guidelines for the disposal of any substance in the land, the identification of the various land; the optimum manner for the utilization of any land, the practices that will conserve the land, the prohibition of practices that will degrade the land.[96]

Basing on the above, NEMA passed the National Environment (minimum standards for management of land quality) requirement of 2001.

The NEA under sections 56 prohibits discharge of hazardous substances, chemicals, oil, etc. into the environment and spiller's liability and Section 57 Prohibits pollution contrary to established principles.

Sections 53 stipulates management of hazardous waste and section 54 illegal traffic in waste. Section 53(2) mandates the Authority to issue guidelines for proper management of wastes. However, according to Section 52 of the National Environment Act and the National Environment (Waste Management) Regulations, the primary responsibility for management of waste lies with the person or company that has generated the waste. These provisions put responsibility and liability at the hands of companies that have generated the waste. This in a certain aspect is proper however NEMA has a responsibility of monitoring to ascertain whether proper mechanisms are implemented to manage the waste. These sections are also general sections on management of wastes but not oil wastes specifically. This Act was enacted before Uganda had prospects in the oil industry therefore its scope towards the industry is minimal.

NEA laid a foundation for other laws such as the Wildlife Act of 1996 meant to promote sustainable management of wildlife, to establish a coordinating, monitoring and supervisory body for that purpose (Uganda Wildlife Authority).

---

[94] Section 6 NEA.
[95] Section 30 (1) NEA.
[96] Section 30 (2) NEA.

The National Forestry and Tree Planting Act of 2003 meant to provide for the conservation, sustainable management and development of forests for the benefit of the people of Uganda and create the national forestry authority (NFA).

4.1.3 Petroleum (Exploration, Development and Production) (PEDP) Act, 2013

The Act is to among others regulate petroleum exploration, development and production, to establish the Petroleum Authority of Uganda, the National Oil Company, to create a conducive environment for the promotion of exploration, to provide for efficient and safe petroleum activities and to provide for the conditions for the restoration of derelict lands.[97]

The PEDP Act under part x creates liability for damage due to pollution.[98] This part makes a general undertaking; it does not specify or even state how oil spills can be avoided. The part only states liability and compensation.

The PEDP further states that:

> The National Environment Management Authority in consultation with the Authority, may grant a license for the management, transportation, storage, treatment or disposal of waste arising out of petroleum activities to an entity contracted by a licensee under subsection (3) on terms and conditions prescribed in the license.

A person contracted by the licensee under subsection (3) shall not carry out those activities without a license issued by the NEMA.[99]

These provisions are properly constituted to oversee waste management at different stages which is a good gesture in that regard more so the penalties the two laws (upstream and down-stream) set once there's a contravention of the sections.

But what has been reported in the Graben is that the current practices are likely to expose the Albertine Graben to more potential risks since a larger area of land in this sensitive eco-system is cleared, dug up and compressed as a methodology to handling the waste in the short run, than if waste was treated and disposed of at once; this was evidenced at Bugungu Waste Consolidation Site where waste had been piled above ground level but had not been properly tucked away.[100]

---

[97]Preamble of the Petroleum Exploration, Development and Production Act (PEDP Act) of 2013.
[98]Section 129 PEDP Act.
[99]Section. 3, 5, 6 PEDP Act.
[100]Report of the Auditor General (2014).

4.1.4 The Petroleum (Refining, Conversion, Transmission and Midstream Storage) (PRCTMS) Act 2013

The Act was intended to inter alia regulate manage, coordinate and monitor midstream operations, to enable the construction, placement and ownership of facilities, to provide for additional and particular health, safety and environment regulations.

The PRCTMS Act directs a licensee or one carrying out midstream operations to comply and take into account environmental principles prescribed by the National Environment Act and other applicable laws.[101]

The Act further obliges a licensee to guarantee that environmental principles and safeguards prescribed under the National Environment Act and other laws applicable are followed while the management of transportation, storage, treatment or disposal of waste arising out of midstream operations is carried out.[102]

The National Environment Management Authority should consult the National petroleum Authority before granting a license for the management of transportation, storage, treatment or disposal of waste arising out of midstream operations to an entity contracted by a licensee,[103] on terms and conditions prescribed in the license.[104]

A person contracted by the licensee under subsection (3) shall not carry out those activities without a license issued by the National Environment Management Authority.

Part IX[105] states liability and compensation wherein a licensee is liable for pollution damage from midstream operations under this Act without regard to fault.[106]

A licensee should endeavor to control the flow and preventing the waste, emission or discharge of petroleum commodities or petroleum products into the environment.[107]

The Act should have exhausted the different strategies to avoid oil spills in the midstream and further state the cleanup strategies incase oil spills occur. Therefore its silence on these pertinent issues is detrimental to land conservation and protection form impacts of the different infrastructure.

---

[101] Section 3 (1) of the Petroleum Refining, Conversion, Transmission and Midstream Storage (PRCTMS) Act 2013.

[102] Section 3(2) PRCTMS Act.

[103] Under Section 3(3) of the PRCTMS Act a licensee shall contract a separate entity to manage the transportation, storage, treatment or disposal of waste arising out of midstream operations.

[104] Section 3(5) PRMCTS Act.

[105] Sections 56 PRCTMS Act.

[106] Section 58 PRMCTS Act.

[107] Section 26 PRMCTS Act.

4.2 Policies

4.2.1 National Environment Policy (NEMP) 1994

The policy objectives include creating sustainable social and economic development, which maintains and enhances environmental quality and resource productivity on a long term basis, which meets the needs of the present generation without compromising the ability of future generations to meet their own needs. It is also meant to collect, analyse, store and disseminate reliable information relating to environmental management issues.

The sustainable development target can be attained with four recommendations from NEMP, these include creation and establishment of appropriate institutional and legal framework, transformation of existing environmental management systems, evolution of a new sustainable conservation culture, revising and modernization of sectoral policies, laws and regulations and establishing an effective monitoring and evaluation system to assess the impact of policies and actions on the environment, the population and economy.

The recommendations are timely since most of the laws were enacted before oil developments were in existence however the policy itself should be upgraded having come into force in 1994, it has been overtaken by several challenges and events such as emerging international best practices, discovery of petroleum in commercial quantities in the Albertine Graben in 2006, climate change impacts such as drought, floods, storms, heat waves and landslides that have had serious effects on agricultural production, food security, incomes, health and livelihoods and technological.

4.2.2 National Oil and Gas Policy (NOGP) of 2008

One of the objectives of the policy is to ensure that oil and gas activities are undertaken in a manner that conserves the environment and biodiversity.

The policy points out that Oil and gas activities in the Albertine can impact the environment in several ways through affecting human beings, wildlife and biodiversity, together with the associated tourism.[108]

Although Uganda is hesitant on joining the EITI to date but the policy calls upon consistency with the internationally recognized Extractive Industry Transparency Initiative (EITI) disclosure standards.

It also ensure that industry activities are undertaken in a manner that conserves forests, biodiversity and biodiversity through strategies such as:

- Strengthening environmental monitoring of oil activities (Strategy b)
- Ensuring that sites at which oil activities are undertaken are restored to original conditions (Strategy d).

---

[108]Section 6.2.4 NOGP.

- Ensuring that oil and gas policies are in harmony with policies for the development and utilization of forest resources, wildlife and tourism.[109]

In collaboration with other policies also supports measures against improper discharge of waste into the natural environment (air, water, land) to ensure safety of animal, fish and human life.

The policy also states the need to minimize the impact footprint in ecologically sensitive sites such as wildlife conservation areas.

For example access roads in such areas should be kept at a minimum and their construction or maintenance should be done in an environmentally responsible manner.[110]

The policy also empowers different institutions to put in place disaster preparedness and response mechanisms for any oil spills in the Albertine Graben, storage facilities and any transportation corridors.[111] The policy further endorses the polluter pays principle through placing responsibility on the licensee to avail mechanisms plus strategies to curb oil spills. In this case the government's role is to design mechanisms of levying penalties for environmental pollution and/or degradation.[112]

5 Conclusion

This chapter reveals that the developments in the Albertine Graben have an impact on the land, several challenges such as oil spillages and waste disposal have been raised as likely impacts to the environment and land. Due to pollution, heavy toxic metals are introduced to the land and later affecting vegetation and plants. It is obviously crucial that laws and environmental policies are improved to protect the Albertine Graben from degradation effects of oil industrial activities. It is also imperative that Environmental guidelines, laws and standards for the petroleum industry in Uganda place liability on the spiller and that the spiller should be responsible for all clean-ups and restoration to the original status of the environment.

5.1 Recommendations

To control oil spills, different measures have been adopted by some of the oil producing countries such as Nigeria where sabotage was criminalized with very

---

[109]Section 7.2.6.1 NOGP.

[110]Section 6.2 NOGP.

[111]Section 6.2 NOGP.

[112]Section 6.2 NOGP.

stiff penalties,[113] creating oil spill funds and further educating the people as to the adverse effects of oil pollution, and also in the government and companies jointly working to ensure that the oil producing communities are provided with basic amenities.

It is pertinent that an oil spill control and prevention law or regulations be passed to guide on different aspects such as the prevention, control and monitoring of oil spill caused by oil plus the establishment of basic principles to be observed in handling oil and other harmful or dangerous substances in facilities, platforms and vessels in Uganda.[114]

The National Environment Management Bill 2017 should be passed to repeal and replace the 1995 National Environment Act.

Part v of the bill stipulates issuing guidelines for the management of the green environment and natural resources by the authority in consultation with the lead agency and includes declaration of special conservation areas by the Minister may, on the advice of the Authority or a lead agency, by statutory instrument,[115] conservation of biological diversity,[116] conservation of biological diversity in situ and compatibility of land use patterns with the conservation of biological diversity[117] and conservation of biological diversity ex situ, especially for species threatened with extinction.[118]

Most of the provisions of the NEA 1995 are outdated to control the several environmental challenges that are existing, emerging international best practices and discovery of petroleum in commercial quantities in the Albertine Graben.

However, there is also need for environmental personnel, expertise and human resources to man the oil industry plus a comprehensive and sustainable effort geared toward sensitizing the general public and raising awareness of environmental problems and solutions available.

Morrison,[119] states that scarcity of funding and meagre budgetary allocations to responsible institutions from both the central and the local governments are major impediments to the effective implementation of environmental laws and policies.[120] Therefore maintenance of a proper environment not only requires laws but also the willingness and commitment from government and other officials.

---

[113]See the 1975 Petroleum Production and Distribution (Anti-Sabotage) Act, Cap 353.

[114]These provisions were proposed under the National Environment (Oil Spill Prevention, Control and Management) Regulations, 2014 that is yet to be passed.

[115]Clause 50 of the National Environment Management Bill.

[116]Clause 58.

[117]Clause 59.

[118]Clause 60.

[119]Rwakakamba (2009).

[120]Rwakakamba (2009).

References

Amnesty International (2009) Petroleum, pollution and poverty in the Niger delta. Available at: https://www.amnesty.org/en/documents/AFR44/017/2009/en/

Bainomugisha A, Kivengyere, Tusasirwe B (2006) Escaping the oil curse and making poverty history: a review of the oil and gas policy and legal framework for Uganda. ACODE Policy Research Series No 20, 2006

Bradbrook AJ, Lyster R et al (2003) The law of energy for sustainable development. IUCN Academy of Environmental Law Research Studies

Centre for Energy, Petroleum and Mineral Law and Policy, CEPMLP (2008) Petroleum transportation and the environment, August 2008. University of Dundee Press, p 44

Ekundayo EO, Obuekwe O (2000) Effects of an oil spill on soil physico-chemical properties of a spill site in a typic udipsamment of the Niger Delta Basin of Nigeria. Environ Monit Assess 60 (2):235–249

Eyinla P, Ukpo J (2006) Nigeria: the travesty of oil and gas wealth. The Catholic Secretariat of Nigeria, Lagos

Global Rights Alert (2017) Global Rights Alert 2017 Update Report. Available at: https://www.globalrightsalert.org/sites/default/files/GRA%20ANNUAL%20REPORT%202017_0.pdf

Golombok R, Jones MI (2015) Oil governance in Uganda and Kenya: a review of efforts to establish baseline indicators on the impact of the oil sector in Uganda and Kenya. UNEP, Nairobi, Kenya. Available at: https://www.macfound.org/media/files/20150730_Oil_Governance_in_Uganda_and_Kenya_Public_Report_FINAL.pdf

Gordon G, Paterson J, Usenmez E (eds) (2011) Oil and gas. Current practices and emerging trends, 2nd edn. Dundee University Press

Government of the Republic of Nigeria (1975) Petroleum Production and Distribution (Anti-Sabotage) 1975 Act (No. 35) (Chapter 353)

Government of the Republic of Nigeria, Ministry of Environment, MOEN (2008) Oil spill logbook investigated

Government of the Republic of Uganda (2008) National Oil and Gas Policy for Uganda. Available at: http://pau.go.ug/uploads/NATIONALOILANDGASPOLICYFORUGANDA.pdf

Government of the Republic of Uganda (2013) Petroleum (Exploration, Development and Production) Act (PEDP Act). Acts Supplement to The Uganda Gazette No. 16 Volume CVI dated April 4th, 2013

Government of the Republic of Uganda (2013) Petroleum (Refining, Conversion, Transmission and Midstream Storage) Act, Acts Supplement to The Uganda Gazette No. 38 Volume CVI dated 26th July, 2013. Available at: https://pau.go.ug/uploads/Petroleum_RCTMS_Act2013.pdf

Government of the Republic of Uganda, National Environmental Management Authority, NEMA (2012) Operational Waste Management Guidelines for Oil and Gas Operations. Available at: https://thisisafrica.files.wordpress.com/2012/07/guidelines-for-oil-and-gas-operations.pdf

Institute for Research in Economics and Business Administration, SNF (2011) Enhancing National Participation in the Oil and Gas Industry in Uganda. SNF-project No 1286: "Local Content in Uganda". The National Content Study in the Oil and Gas Sector in Uganda, Bergen 2011. Available at: https://www.snf.no/Files/Filer/Publications/SNF-R13_11.pdf

International Monetary Fund (2010) Annual Report 2010. Supporting a balanced global recovery. Available at: https://www.imf.org/~/media/Websites/IMF/imported-flagship-issues/external/pubs/ft/ar/2010/eng/pdf/_ar10engpdf.ashx

Johnson D et al (2014) Fuelling conflict or development? Cross-border oil & gas resources in the African Great Lakes. Pole Institute, Dossier January 2014. Available at: http://www.pole-institute.org/sites/default/files/FUELLING%20CONFLICT.pdf

Kakura K, Ssekyana I (eds) (2009) Handbook of environmental law in Uganda, vol 1, 2nd edn

Kasimbazi EB (2012) Environmental regulation of oil and gas exploration and production. J Energy Nat Resour Law 30(2)

Laitos JG, Toumain JP (1992) Energy and natural resources law in a nutshell. West Group

Leopold A (1989) A Sand County Almanac, and sketches here and there. Oxford University Press, USA

Ling QC, Yan S, Bao ZY (2007) Metals contamination in soils and vegetables in metal smelter contaminated sites in Huangshi, China. Bull Environ Contam Toxicol 79(4):361–366

Ministry of Energy and Mineral Development of the Republic of Uganda, MEMD (2017) National Content Policy for the Petroleum Subsector in Uganda. Available at: http://pau.go.ug/uploads/ NATIONAL_LOCAL_CONTENT_POLICY_FOR_PETROLEUM_IN_UGANDA.pdf

Mubiru DN, Namakula J et al (2017) Conservation farming and changing climate: more beneficial than conventional methods for degraded Ugandan soils. Sustainability 9:1084

National Environmental Management Authority, NEMA (2009) Environmental Sensitivity Atlas for the Albertine Graben. Available at: http://chein.nema.go.ug/wp/?wpfb_dl=46

Netherlands Commission for the Environmental Assessment, NCEA (2012) Strategic Environmental Assessment of oil and gas activities in Albertine Graben, Uganda. Available at: http://www. commissiemer.nl/docs/cms/Key%20sheet%20SEA%20Albertine%20Graben,%20Uganda.pdf

Nkonya EM, Pender JP et al (2002) Development pathways and land management in Uganda. Paper presented at the Conference on Policies for Land Management in the East African Highlands, United Nations Economic Commission for Africa (ECA), Addis Ababa, Ethiopia, April 24–26, 2002

Office of the Auditor General for the Republic of Uganda (2014) Annual Report of the Auditor General for the Year ended June 30th, 2014. Available at: http://www.oag.go.ug/4961-2/

Oil in Uganda (March 6th, 2018) Buliisa residents demand affirmative action on land and property ownership. Available at: http://oilinuganda.org/features/land/buliisa-residents-demand-affirmative-action-on-land-and-property-ownership/

Oxford Institute for Energy Studies (2015) Oil in Uganda: hard bargaining and complex politics in East Africa. October 2015. Available at: https://www.oxfordenergy.org/wpcms/wp-content/uploads/2015/10/WPM-601.pdf

Plumptre AJ, Mutungire N et al (2009) Biodiversity Surveys of Kabwoya Wildlife Reserve and Kaiso Tonya Community Wildlife Area. Available at: https://albertinerift.wcs.org/DesktopModules/Bring2mind/DMX/Download.aspx?EntryId=11606&PortalId=49&DownloadMethod=attachment

Rwakakamba TM (2009) How effective are Uganda's environmental policies? Mt Res Dev 29 (2):121–127

Stevens L (2014) Visualising spill risk: understanding and assessing regions of heightened vulnerability associated with increased seaborne transport of oil. Int Oil Spill Conf Proc 2014 (1):300151

The Independent (July 18th, 2017) OIL & GAS: Uganda unveils 2017 National Suppliers' Database. Available at: https://www.independent.co.ug/uganda-unveils-2017-national-suppliers-database-oil-gas-industry/

The Legal Framework for Sand Mining in Uganda

Tajudeen Sanni

1 Introduction

A screaming headline in the London Guardian newspaper reads: "Sand mining: the global environmental crisis you've probably never heard of".[1] The headline captures the problems associated with mining of sand in many countries around the world.

This is by no means inexplicable: the global urbanization boom is consuming large amounts of sand as the major component of concrete and asphalt used in construction works. Mining of tracts of land and water bodies for sand to meet increasing demand for sand in the construction industry has thus been on the increase globally. It has been so rampart that the issue was presented, in 2012, as a serious international threat to biodiversity at the Conference of Parties of Convention on Biodiversity in Hyderabad.[2]

Concerns have been raised not only about the impact of sand mining on the environment or biodiversity but on agricultural productivity in general.[3] From Uganda to India to United States, the problem associated with this variant of mining

---

[1]The Guardian (February 27th, 2017).

[2]In 2013 the Awaaz Foundation and the Bombay Natural History Society presented the issue of sand mining to the Secretariat of Convention on Biodiversity as a major international threat to coastal biodiversity.

[3]Agriculture contributes about 26% to Uganda's GDP. It is the largest source of export earnings, about 53% of export earnings from 2007 to 2012/13. Over 75% of the labour force are employed in the sector. See Trading Economics (2019).

T. Sanni (✉)
Kampala International University, School of Law, Kampala, Uganda

Nelson Mandela University, Department of Public Law, Port Elizabeth, South Africa

© Springer Nature Switzerland AG 2020
H. Yahyah et al. (eds.), *Legal Instruments for Sustainable Soil Management in Africa*, International Yearbook of Soil Law and Policy,
https://doi.org/10.1007/978-3-030-36004-7_6

is a cause of concern.[4] According to the Uganda National Agriculture Policy, sustainable use and management of water, soil and land resources remains a critical factor for agricultural production and productivity.[5] To that effect, one operational principle of the Policy is the requirement that the Government ensures key agricultural resources including soils and sand for agricultural production are sustainably used and managed to enhance adequate output for the current and future generations.[6] While there has been an effort to apply this principle in general, the extent of its application in the particular context of sand mining is an important question.

2 Background

In recent years, in Uganda, there has been phenomenal increase in sand mining activities which is a threat to sustainable management of soil as envisaged in a number of environmental instruments. This has to a large extent contributed to land degradation, desertification and the destruction of important economic trees most especially of the indigenous variety.

In spite of prevalence and impact on the environment and agricultural lands, sand mining has not been given the attention it deserves. While some studies have been conducted on the impacts of sand mining, the legal aspect has rarely been a subject of extensive scrutiny. Recently, attention was momentarily focused on mining along river banks and lake shores[7]; sand mining on shore or on land has not generally been

---

[4]In the case of United States, for example, the industry has been growing by nearly 10% annually since 2005 due to its of hydrocarbon extraction. Much of the market size for mining is held by Texas and Illinois. Silica sand mining business has more than doubled since 2009 on account for the need for this particular type of sand, which is used in a process known as hydraulic fracturing. Wisconsin is one of the five states which produce nearly 2/3 of the nation's silica. As of 2009, Wisconsin, along with other northern states, is facing an industrial mining boom, being dubbed the "sand rush" on account of the new demand from large oil companies for silica sand. According to Minnesota Public Radio, "one of the industry's major players, U.S. Silica, says its sand sales tied to hydraulic fracturing nearly doubled to $70 million from 2009 to 2010 and raked in nearly $70 million in the first nine months of 2011." See MPR News (March 8th, 2012).

[5]National Agriculture Policy, NAP (2013), Par 2.6.1.

[6]NAP, Par 3.1.

[7]The Ugandan National Environment Management Authority (NEMA) issued a warning of devastating environmental consequences if Chinese investors insist on mining sand in Lake Victoria. NEMA said Mango Tree Group Ltd was only permitted to manufacture ships and must first seek approval to engage in sand mining. "We want to categorically state that Mango Tree Group Ltd's activities of engaging in sand mining are illegal. We only cleared them [Mango Tree Group Ltd] to manufacture ships after they secured a license from the Uganda Investment Authority, not carrying out dredging on the lake shores," said NEMA. "The sites where they want to excavate sand are very crucial. We warned them against tampering with them because any destruction may spark off a serious ecological disaster we may not handle as a country. They can think of other areas but an EIA [Environmental Impact Assessment] has to first be carried out," she added. Background: On September 2 last year, the company submitted its first EIA report to NEMA, seeking approval to

considered particularly in the context of impacts on farm lands. In view of this, the study seeks to consider the legal aspects of sand mining in Uganda. The primary objective of this study is to carry out a legal analysis of laws relating to sand mining in Uganda. Specifically, the chapter will analyze relevant provisions in the law as it stands now. To that effect, recommendation is made to improve on the laws governing the mining of sands in the country.

3   The Meaning of Sand Mining

It is apposite starting an attempt to understand sand mining with what sand means both in itself and in relation to soil which is an important part of the lithosphere.[8] In one sense, sand is defined as a loose material that consists of broken rocks or mineral grains. In another sense, the word sand is used to describe a specific category of soil which contains more than eighty five percent of sand-sized particles. The word soil is derived from a Latin word *solum* which means earthy material.[9] In agricultural terms, soil is simply the top layer of the earth which supports plant life.[10] It is the top layer of the land surface of the earth that is made of smashed rock particles, water, humus, and air.

While both can be distinguished in the first sense above described, in the second sense sand is characterized as a type of soil. For the purpose of this work, sand is used in the two senses as both are important for agricultural productivity.[11]

Sand mining is a multi-billion dollar industry selling at an average of $90 per cubic yard.[12] Sand mining is the extraction of sand normally through an open pit and from beaches and inland dunes as well as from sea, lake and river beds. Sand is used in manufacturing in a number of ways, for example, as an abrasive or in concrete. It is also used on icy and snowy roads, usually mixed with salt, to lower the melting point temperature on the road surface. Sand can replace eroded coastline. Sand is

---

excavate sand from three sites on the lake shores of Nkumba near Entebbe, near Kimi Island in Mukono District and near Kavejanja-Buusi Island, Wakiso District. But NEMA rejected the EIA on the basis that the company's activities would have a negative impact on the marine ecology. In a letter to Mango Tree Group directors dated 1st June 2017, Dr Tom Okurut, the NEMA executive director, explained that Kimi Island, for instance, is a tilapia breeding area, Buusi and Nkumba are breeding and nursery areas for Nile Perch as well as active fishing grounds, which must not be tampered with. See AllAfrica (February 2nd, 2018).

[8]Lithosphere refers to the solid component of the earth and includes the crust mantle as well as outer and inner core. See Prabakhar (2000), pp. 125–126.

[9]Prabakhar (2000), p. 126.

[10]Rajan and Rao (2007), p. 1.

[11]The mineral part of soil is divided into three particle-size categories: sand, silt, and clay. It should be noted that sand, clay and silt are collectively characterised as the fine earth fraction of soil: they are 2 mm in diameter.

[12]According to one estimate, it is worth $70 billion. See Mills and Staats Shrinking Shores: Florida sand shortage leaves beaches in lurch. See Naples News (January 13th, 2017).

also mined to extract rutile, ilmenite and zircon containing essential elements titanium and zirconium which normally occur combined with ordinary sand and have to be separated in water by virtue of their different densities, before the sand is redeposited.

In considering remedial legal measure on sand mining, it is worth noting that land which was originally used for grazing and cultivation has been reduced to pits and ponds that are a threat to humans and cattle, apart from being unsuitable for cultivation. According to statistics from the lands ministry, over 2000 housing units are set up annually. The construction industry constitutes over twelve percent of Uganda's Gross Domestic Product (GDP) and has witnessed significant surge in the last twenty years. In spite of upsurge in inflation, the sector has remained on a steady path of growth and development. It is this sector, in addition to industrial sector—for example the industrial subsector involved in glass-making—that has contributed significantly to the development of the sand economy with all its attendant effects including on farmlands. It is therefore against the above background that the legal regime is examined.

4 Effects of Sand Mining

Sand mining has got a number of advantages: providing an essential raw material for infrastructural development for example road and housing works, providing raw materials for industrial products such as glass-making, serving as a source of employment for sand miners among other economic advantages. The advantages need to be contextualized against its negative effects so as to provide a basis for a strategic approach for dealing with it.

Sustainable development is one of the fundamental principles of natural resources management in Uganda; economic output needs to be considered with environmental implication.

This is even more apposite considering the fact that the costs of natural resource degradation in the country stands at no less than seventeenth percent of GDP per year.[13]

Commercial excavation of sand is lucrative because of the purposes sand serves in different aspects of the economy including construction as earlier pointed out. Silica sand, quartz (a hard white mineral consisting of silicon dioxide) that overtime, through the work of water and wind, has been broken down into tiny granules is one of the most common types and has many utilities including making glass. Despite these advantages, sand mining activities have had adverse impacts and caused negative effects on the environment in a number of places such as Lwera wetland

---

[13]NEMA (2016).

including water pollution, destruction of the vegetation and excavation of pits and trenches which destroys the aesthetic beauty of the landscape.[14]

The Uganda Land Policy identifies different causes of soil degradation such as inappropriate use of arable land including excessive cultivation, cropping too frequently, inadequate fertilization or baring of the soils through grazing or removal of crop residues.[15] Sand mining does not feature conspicuously as one of them in spite of scientifically documented evidence to that effect. As it were, sand mining along the Kampala–Masaka highway, for example, is causing the silting of Lake Victoria thus depriving fish and other aquatic lives of breeding grounds. In the same vein, sand mining is a direct cause of erosion, and impacts the local wildlife.[16] Various animals depend on sandy beaches for nesting clutches, and mining has led to the near extinction of gharials (a species of crocodile). Disturbance of underwater and coastal sand causes turbidity in the water, which is harmful for organisms like coral that need sunlight for their survival. It can also destroy fisheries, causing financial loses to fish farmers.

Destruction of physical coastal barriers, such as dunes, sometimes leads to flooding of beachside communities, and the destruction of picturesque beaches causes tourism to suffer incalculable set back. Sand mining is regulated by law in many places, but is often done illegally. Globally, it is a $70 billion industry, with sand selling at up to $90 per cubic yard.

This, it must be reemphasised, has to a large extent contributed to land degradation and desertification through the destruction of economically important trees, mostly indigenous in nature. The practice leaves in trail bare soil and a large expanse of gullies which can collect water during rainy seasons. This can result not only in health-related problems for local communities, but can cause negative impacts on the environment as well as agricultural productivity. More specifically such effects include loss/reduction of farmlands and grazing lands, source of breeding grounds for mosquitoes and spread of diseases, erosion, loss of vegetation or fertility, loss of biodiversity and economically important trees, destruction of landscape and beauty, sand and dust pollution and pollution of underground water.[17] In the US, for example, the Wisconsin Department of Natural Resource (WDNR) released in 2012 a final report on the silica sand mining in Wisconsin titled Silica Sand Mining in Wisconsin.[18] The recent boom in silica sand mining has caused concern from residents in Wisconsin on issues that include quality of life issues and the threat of silicosis. According to the WDNR these issues include noise, lights, hours of operation, damage and excessive wear to roads from trucking traffic, public safety concerns

---

[14]Daily Monitor (September 30th, 2016).

[15]Uganda Land Policy (2013), Par 1.4.1.

[16]According to Ministry of Animal Industry, Agriculture and Fisheries, 46% of Uganda's soil is already degraded with 10% being highly degraded. See MAAF (2019).

[17]Saviour (2012), pp. 125, 131.

[18]According to the Wisconsin Department of Natural Resources Report (2017), there are currently no less than thirty-four active mines and twenty-five mines in development in Wisconsin.

from the volume of truck traffic, possible damage and annoyance resulting from blasting, and concerns regarding aesthetics and land use changes. The situation in Uganda is not different. By far the greatest impact of mining on the soil resources is due to opencast mining, which is causing tremendous deterioration of soil quality.[19] This has negative effects on productivity of the soil for farming purposes. Similarly surface mining, as contrasted with underground mining, results in the destruction of the existing vegetation and soil profile. Removal of overburden and waste rock and its replacement in waste dumps or the mined-out pit have the potential to significantly transform the topography and stability of the landscape.[20] Land refers to soil, landforms, geology, climate and hydrology, the plant cover, and fauna including insects and microorganisms. The nature of utilization under which it is currently put or the possible kinds of uses under consideration for the future is referred to as Land Use and is no doubt under threat from sand mining activities.[21]

The Lake Victoria Basin is endowed with alluvial depositions that contain sand, which is a highly sought after and widely used resource by the construction industry. The deposit of sand in the basin is deeper in the Western part of Uganda, especially in the vicinity of Lwera and Bukatata. It is therefore not surprising that out of all permitted sand mining projects, over eighty percent of them are based in the West of the Lake Victoria basin, particularly in Lwera area.

In Lwera, commercial sand mines started recently, and their emergence over the years is wholly attributed to a growing demand for clean sand. The situation contrasts with some years back when sand was mined using simple local tools, such as hoes and spades. During that time, large scale mining in Uganda was confined to few places such as Bukakata, in the 1960s, where sand was excavated to be used for glass making by the East African Glass Works Limited. As will be seen anon in the next section, sand mining activities come with negative impacts, and Lwera has not been spared. Kamaliba, a traditional fishing village, which is surrounded by three mining companies, has also been negatively affected by the activities of sand miners. This has led to massive loss of shelter, toilets, access roads, recreational space and land for cultivation, as some of the developers have expanded their mines beyond the permitted boundaries.

5   The Legal Regime

The law describes soil to include sand. Under the National Environment Act "soil" includes earth, sand, rock, shale, minerals, vegetation, and the soil flora and fauna in the soil and derivatives thereof such as dust. This means that soil includes sand.

---

[19]Wisconsin Department of Natural Resources Report (2017).

[20]Wisconsin Department of Natural Resources Report (2017).

[21]Uganda Land Policy Preamble.

It is the responsibility of the State, as the custodian and trustee of natural resources,[22] to protect the environment from degradation of whatever type. In terms of the Constitution, the State is mandated to promote sustainable development and public awareness of the need to manage land, air, water resources in a balanced and sustainable means for the present and future generations.[23] Pursuant to this, the government is to make laws and policies to regulate the use of land.[24]

Under Ugandan laws, sand is considered a mineral resource and generally subject of mineral right as provided for under the Mining Act.[25] In terms of the Mining Act, mineral "means any substance, whether in solid, liquid or gaseous form occurring naturally in or on the earth, formed by or subject to a geological process, but does not include petroleum, as defined in the Petroleum (Exploration and Production) Act, 1985, water or building mineral".[26]

Sand depending on how it is characterized falls under building mineral or industrial mineral. Building mineral refers to "rock, clay, gravel, laterite, murram, sand, sandstone and slate, which is mined by a person from land owned or lawfully occupied by him or her for his or her own domestic use in Uganda for building, or mined by a person for his or her own use for road-making, and includes such other minerals as the Minister may from time to time declare by notice published in the Gazette, to be building minerals."[27] Article 244(1)(3) of the constitution refers to this category: they are not to be considered as minerals for the purpose of the provisions in that part of the Constitution (namely ART, 244).

A number of important and relevant points flow from the above provision. First, sand is generally not considered a mineral in that particular provision. While in practice this is strictly observed and sand is never treated as a mineral, the above provision's emphasis on 'own domestic use' leaves a gap that appears to exclude commercial sand mining. If the provision is read in the context of other provisions of the Act and other laws, can that position legally hold water under all circumstances as is generally believed? A holistic reading of the Act does suggest that commercially mined sand enjoys classification as a mineral resource. To this effect, consider definition of industrial mineral. According to the Act, industrial minerals "means barite, rock, clay, dolomite, feldspar, granite, gravel, gypsum, laterite, limestone, mica, magnesite, marble, phosphate rock, sand, sandstone, slate and talc, which is commercially mined by a person for use in Uganda or industrially processed into finished or semi-finished products, and includes such other minerals as the Minister may from time to time declare by notice published in the Gazette, to be industrial minerals".[28] That definition includes sand which is commercially mined. When this

---

[22]See Constitution Art. 26, see also Section 44 Land Act.

[23]Constitution of Republic of Uganda, Directive Principles of State Policy xxvii(i).

[24]Art. 242 of the Constitution of Republic of Uganda.

[25]Mining Act (2003).

[26]See Sec. 2 Mining Act.

[27]Sec. 2 Mining Act.

[28]Sec. 1 Mining Act.

provision is placed in the context of the definition of minerals under the Act, the status of sand as a mineral, if it fulfills statutory requirement to that effect puts it in similar category with what the Act calls precious minerals.[29] Repeated for emphasis "mineral" means any substance, whether in solid, liquid or gaseous form occurring naturally in or on the earth, formed by or subject to a geological process, but does not include petroleum, as defined in the Petroleum (Exploration and Production) Act, 1985, water or building mineral."[30] The definition only excludes petroleum,[31] water or building materials. It does not exclude what the Act calls industrial minerals including sands which are commercially mined for use or industrially processed into finished or semi-finished products.

The basic difference between building mineral or sand on one hand and industrial mineral or sand on the other hand is that the former is for domestic or noncommercial use while the latter is for commercial or industrial use. In the latter case it must be one "commercially mined by a person for use in Uganda or industrially processed into finished or semi-finished products" Sand rich in silica is used for making glasses, that is an example of industrial use, sand excavated for sale to be used for construction work, that is a commercial use. Sand mined in both instances and other similar cases is thus classified as industrial mineral.

Where sand is mined for the purpose of extracting minerals that are classified under the Mining Act as precious minerals, would such fall under industrial minerals, building minerals or precious minerals categories? Precious minerals can either be precious stones or precious metals.[32] Precious stones include agate, amber, amethyst, cat's eye, chrysolite, diamond, emerald, garnet, opal, ruby, sapphire, turquoise and all other substances of a similar nature to any of them.[33] Precious metals include gold, silver, platinum, iridium, osmium, palladium, ruthenium, rhodium, or any other rare earth elements.[34] It is submitted that where sand is mined for the purpose of eventually extracting any of the above mentioned minerals or minerals of similar nature with them; it would be neither building minerals nor even industrial minerals as defined in the Act and as described above. The operative words for "building mineral" imply that rock, clay, gravel, laterite, murram, sand, sandstone and slate themselves are put to own use or meant for domestic use.[35] Similarly, in the description of "industrial minerals" the operative words imply that

---

[29]Precious minerals are categorised into previous stone and precious metals. According to the Act "precious minerals" include—(i) precious stones, namely agate, amber, amethyst, cat's eye, chrysolite, diamond, emerald, garnet, opal, ruby, sapphire, turquoise and all other substances of a similar nature to any of them; and (ii) precious metals, namely gold, silver, platinum, iridium, osmium, palladium, ruthenium, rhodium, or any other rare earth elements" See Sec 2 Mining Act.

[30]Sec. 2 Mining Act.

[31]Petroleum is defined under the Petroleum (Exploration, Development and Production) Act, PEDP Act (2013).

[32]Sec. 2 Mining Act.

[33]Sec. 2 Mining Act.

[34]Sec. 2 Mining Act.

[35]Sec. 2 Mining Act.

barite, rock, clay, dolomite, feldspar, granite, gravel, gypsum, laterite, limestone, mica, magnesite, marble, phosphate rock, sand, sandstone, slate and talc are themselves used and are commercially mined or industrially processed.

The implication of the legal characterization of the types of sand mining that falls under industrial minerals regime (or precious minerals regime) is that the specific rules governing mining under the Act apply to them *mutantis mutandis*.[36] This translates to mean that miners of sand at industrial or commercial levels are statutorily required to pay royalties contrary to what obtains at the present.[37]

It is provided that all minerals mined in the course of prospecting, exploration, mining or mineral beneficiation operations shall be subject to the payment of royalties on their gross value which is calculated on the basis of the prevailing market price of the minerals at such rates as shall be prescribed by the relevant authority.[38] The owner of such lands is entitled to such royalties which is not being paid under the current dispensation.

In terms of the Act

> The owner or lawful occupier of any land subject to a mineral right is entitled to compensation under either section 82 of this Act or to a share of royalties under section 98 of this Act.[39]

It should be noted that the laws governing environmental sustainability in the mineral sector also applies to this category of mining namely sand mining. The relevant provisions are discussed in Environmental Sustainability and Sand Mining under three subheadings to wit: Environmental Restrictions and Sand Mining, Environmental Impact Assessment in the Sand mining Industry and Soil Standards vis-à-vis Sand mining.

6 Sand Mining and Environmental Sustainability

Apart from the provisions of National Environment Act, there are specific provisions in the Mining Act for environmental protection in the mining sector. Such provisions offer special protection to the environment including the soil in the face of mining activities. Neither is a mining lease to be granted nor renewed without taking into cognizance environmental implications thereof.[40]

---

[36]One of such rules in the case precious minerals is that relating to the Gold Smith Licence. The Act provides that no person shall manufacture any article from any precious mineral or from any substance containing any precious mineral without obtaining a licence to that effect called goldsmith's licence. See Sec. 73 and 74 Mining Act.

[37]Sec. 98 Mining Act.

[38]Sec. 98 Mining Act. The section is subject to sec 100 of the Act.

[39]See Sec. 83 Mining Act. Section 82 provides for compensation in general while section 98 is about royalties.

[40]See Secs. 43(3)(b) and 47 of the Mining Act.

6.1   Sand Mining Versus Environmental Restrictions

Placing restriction or limits on mining activities in environmentally sensitive or vulnerable areas is one of the tools employed in the Mining Act for environmental protection. In general, the rights conferred by a mineral right are required to be exercised reasonably and in such a manner as not to adversely affect the interests of any owner or occupier of the land on which the rights are exercised.

Specifically, restrictions have been placed on the exercise of mineral rights within specified parameters of land described as including "land beneath any water, the seabed and subsoil of such land."[41]

For example, in terms of the Act, the holder of a mineral right shall not exercise any of his or her rights under that mineral right in a number of places including an area within at least five meters or a greater distance as may be prescribed, of land which has been cleared or ploughed or otherwise prepared in good faith for the growing of, or upon which there are growing agricultural crops.[42] This means that holder of commercial or industrial sand mining right duly conferred on him under the Act cannot exercise that right of mining, all things being equal, in an area within at least five meters of a piece of land on which agricultural crops are growing or a piece of land prepared for the growing of agricultural product. The same applies to a piece of land which is the site of or which is within the perimeters of a hundred meters, or such greater distance as may be prescribed, of any cattle dip, tank, or similar body of water, except with the written consent of the owner or lawful occupier or the duly authorized agent of the owner or lawful occupier of that piece of land.[43]

Where industrial or commercial sand-mining activity is to be carried out within two hundred meters from a lake or within one hundred meters from any rivers, it must be pursuant to a permit issued under the National Environment Act.[44] This means that industrial or commercial sand mining activities in lakes and rivers as well as on river banks and lake shores require a permit from the National Environmental Management Authority (NEMA).This is line with the mandate of NEMA to take measures, in consultation with relevant agency, for the protection of river banks and lake shores from human activities that will adversely affect these water bodies.[45] As a general rule excavation or drilling of rivers and lakes whether for sand or otherwise is prohibited except with the permission of NEMA and relevant agency.[46] A person who intends to carry out excavation or drilling of a river bank or lake shore shall seek permission by way of application to the Executive Director of NEMA.[47]

---

[41]Sec. 2 Mining Act.

[42]Sec. 72(b)(i) Mining Act.

[43]Sec. 78 (1)(b)(ii) Mining Act.

[44]This is in terms of the implication of Section 78(1)e) Mining Act.

[45]See sec 35(i) National Environment Act (NEA).

[46]Sec. 34(1)(b) and 34(2) NEA.

[47]See Regulation 23(1)(b) of The National Environment (Wetland, Riverbanks and Lake Shores) Management Regulations (2000). The application must be done using the form set out in the First

Generally, an application for a mineral right must show whether the mining activity will use water for prospecting, exploration and mining operations within the defined boundaries of his or her mineral right or if any natural source of water will be used for washing product at the site.[48] It must also be indicated that the applicant for licence intends to convey specified water to the area of his or her mineral right from any natural water supply outside the boundaries of the mineral right required for the relevant operations or wants to occupy any land that may be required for the construction of a dam, reservoir or pumping station and for the conveyance of such water to the area where the water is used, by means of pipes, duets, flumes, furrows or otherwise, and for such conveyance to have a right of passageway.[49] Where he decides to construct any works that will be used for the collection, storage or conveyance of such water, he is also required to so indicate in the application.[50] In all cases relating to acquisition of the right to use water in any manner particularly for the above purposes, Water Act, 1995, is the applicable law.[51] The overarching provision is that rights in wetlands and in the waters of any spring, stream, river, watercourse, pond or lake on or under public land, are vested in the Government; and no such wetlands or water shall be obstructed, dammed, diverted, polluted or otherwise interfered with, directly or indirectly, except in accordance with the Water Act.[52]

Where such industrial or commercial sand mining is to be carried out in a national park or game reserve, written permission of the authority having control of the park or game reserve must be sought.[53] The relevant authority in charge of wildlife conservation areas such as game reserves and national parks is the National Wildlife Authority whose statutory functions include controlling and monitoring industrial and mining activities in wildlife protected areas.[54] In instances where such mining activities take place within a forest reserve, it must be with the consent of the body in charge of forestry.[55]

Generally, the holder of a mining lease may, if he or she requires the exclusive use of the whole or any part of the mining area concerned, and if so requested by the

---

Schedule to this regulation. Same explicit provisions are not made in respect of wetland. However, pursuant to regulation 11 of the Regulations, the second schedule to the Regulations lists exploitative commercial activities on wetland as one of the regulated activities on wetlands.

[48]Sec. 87(a-b) Mining Act.

[49]Sec. 87(c-d) Mining Act.

[50]Sec. 87(e) Mining Act.

[51]Sec. 87(2) Mining Act.

[52]See Sec. 86 Mining Act.

[53]This is by the implication of Section 78(1)(g) Mining Act.

[54]See Sec. 5(h) Uganda Wildlife Act.

[55]See Sec. 78(1) h) Uganda Wildlife Act. The consent referred to here and in other instances earlier discussed must be made in line with the relevant instruments of the agencies concerned such as National Forestry Authority in this instance. See Sec. 78 (2) which provides: "Any consent under subsection (1) of this section may be given subject to such conditions as are specified in the instrument of consent."

owner or lawful occupier of any part of such area, obtain a land lease or other rights to use the area upon such terms as to duration or the extent of the affected land, as may be agreed between the holder and the owner or lawful occupier of the land in question, or failing an agreement, as may be determined by arbitration.[56] In assessing any rent payable under this section, an arbitrator shall determine the rent in relation to values, at the time of arbitration, current in the area in which the mining lease is situated, for land of a similar nature, but without taking into account any enhanced value due to the presence of minerals.[57]

The mineral right holder is required, on demand made by the owner or lawful occupier of any land subject to such mineral right, to pay the owner or lawful occupier fair and reasonable compensation for any disturbance of the rights of such owner or occupier; and for any damage done to the surface of the land by the holder's operations.[58] A licensee shall on demand made by the owner of any crops, trees, buildings or works damaged during the course of such operations, also make payment in compensation for any crops, trees, buildings or works that have been damaged in the process.[59] The implication of the above is that where the environmental restrictions applicable to minerals are applied to sand mining being a class of minerals, it helps to environmentally secure the land from degradation that may affect it productivity.

6.2   Sand Mining and Environmental Impact Assessment

The clearest specific reference in the National Environment Act relating to sand mining is the statutory requirement for Environmental Impact Assessment before engaging in excavation of sand. The third schedule to the Act is a list of activities for which Environmental Impact Assessment is required. On the list is "mining, including quarrying and open-cast extraction of—(a) precious metals; (b) diamonds; (c) metalliferous ores; (d) coal; (e) phosphates; (f) limestone and dolomite; (g) stone and slate; (h) aggregates, sand and gravel; (i) clay; (j) exploration for the production of petroleum in any form".[60] The schedule is made pursuant to section

---

[56]Sec. 81(1).

[57]Sec. 81(2).

[58]See Sec. 82(1).

[59]The law as per section 81(1)(I) further provides that in "assessing compensation payable under this section, account shall be taken of any improvement effected by the holder of the mineral right or by his or her predecessor in title the benefit of which has or will accrue to the owner or lawful occupier of the land; (ii) the basis upon which compensation shall be payable for damage to the surface of any land shall be the extent to which the market value of the land upon which the damage occurred has been reduced by reason of the damage; (iii) no compensation shall be payable to the occupier of a state grant land in respect of any operations under a mineral right existing at the date of such state grant".

[60]Third Schedule, Item 6.

19 of the Act which mandates carrying out Environmental Impact Assessment for any activity that will or is likely to have effect on the environment. The section apparently applies to mining of sand for whatever purposes as long as the mining activity will or is likely to affect the environment as envisaged under the National Environment Act.

Accordingly every holder of a mining exploration license or a mining lease is required to carry out an environmental impact assessment of his or her proposed operations.[61] The programme of proposed mining operations takes proper account of Environmental Impact Assessment, Environmental Impact Research, Environmental Statement and safety factors. The holder of a license referred to in subsection (1) of this section shall commence his or her operations under this Act only after securing a certificate of approval of his or her proposed operations from the National Environment Management Authority. The holder of a licence referred to in subsection (1) of this section shall carry out an annual environmental audit, and shall keep records describing how far the operations conform to the approved environmental impact assessment. The provision of subsections (1) and (3) of this section relating to environmental impact assessment and audit shall not apply to the holder of a location license.

6.3   Sand Mining and Soil Standards

A raft of soil standards is one of the safest safeguards for the protection of the quality of sand for agricultural purposes.[62] Good soil has been described as the most valuable asset a nation can have.[63] It is no wonder therefore that the responsibility of National Environmental Management Authority includes liaising with relevant agency to establish minimum standards to maintain the quality of the soil.[64] For that purpose, the authority is mandated to issue guidelines for the disposal of any substance in the soil, the identification of the various soils, the manner for the utilization of any soil and the practices that will conserve the soil.[65] It is also the responsibility of the authority to issue guideline that will prohibit practices capable of degrading the soil.[66]

The guidelines would cover all practices, including sand mining, by necessary implication, particularly in terms of standards that help determine the benchmarks for soil quality as a way of ensuring that no sand excavation activity degrades the

---

[61] Section 108 (NEA).

[62] In terms of Section 2 NEA, "standard" means the limits of pollution established under Part VI of this Act or under regulations made under this Act or any other law".

[63] Prabakhar (2000), p. 126.

[64] Sec. 30 (1)(b) NEA.

[65] Sec. 30(2)a-d NEA.

[66] Sec. 30(2)e NEA.

soil. Pursuant to these provisions, the authority promulgated the National Environment (Minimum Standards for Management of Soil Quality) Regulation, 2001.The purpose of the regulation is to establish minimum soil quality standards for maintaining, restoring and enhancing the inherent productivity of the soil, prescribe minimum standards for specific agricultural practices, establish criteria and procedures for benchmarking and measuring soil quality as well as prescribing guidelines for managing the soil.[67] The regulation prescribes different soil standards for different soil types and agricultural activities spelt out in the schedules to the regulation. It provides that no person shall use land in contravention of the soil quality parameters established under the regulations.[68]

For sand mining, this translates to mean that it must be carried out without contravening the parameters set out in the regulation. The ambit of the regulation is all encompassing. Specifically, the Mining Act requires mining activities to be conducted with due regards to and in accordance with the standards and guidelines prescribed under the National Environment Act.[69] However, the holder of an exploration license or a mining lease may exceed the standards and guidelines prescribed under the National Environment Act, 1995, if he has authorization by way of a pollution license issued under the National Environment Act.[70] No specific requirement is made for Environment Impact Assessment for this purpose nor for public involvement though there are general provisions in the Act for impact assessment and public participation which the environmental authority may invoke when it deems necessary to protect the environment.

7   Institutional Arrangements

Soil and Water Conservation and Water for Agricultural Production are under the Department of infrastructural development (Farm Development) of the Ministry of Agriculture Animal Industry & Fisheries.[71] Government ministries and government

---

[67]See Regulation 3 of the National Environment (Minimum Standards for Management of Soil Quality) Regulations, 2001.

[68]See Regulation 11.

[69]Sec. 109 NEA. The relevant section reads: "Environmental protection standards (1) There shall be included in every exploration licence or mining lease granted under this Act a condition that the holder of such licence or lease takes all necessary steps to ensure the prevention and minimisation of pollution of the environment in accordance with the standards and guide lines prescribed under the National Environment Statute, 1995".

[70]Sec. 109(2) NEA.

[71]Ministry of Agriculture, Animal Industry and Fisheries is a Government Ministry charged with creating an enabling environment in the Agricultural Sector. It is commonly known as Ministry of Agriculture and carries out its role by enhancing crop production, improving food and nutrition security, widening export base and improved incomes of the farmers. The Ministry has Directorates which include Animal Resources, Crop Resources, and Departments for Planning, Finance and Administration.

agencies such as Ministry of Water & Environment, Ministry of Energy & Mineral Development, NEMA, etc. have statutory roles relating to sand mining. The National Environment Authority is required by law to establish criteria and procedures for the measurement and determination of soil standards and the minimum standards for the management of soil quality.[72] The authority is required to carry out this function in consultation with the lead agency. However, the lead agency is neither categorically specified by the National Environment Act nor by any other law. It would seem legally plausible that this role is performed by such bodies as the relevant department in the Ministry of Agriculture particularly Soil and Water Conservation Department though this department is not an independent agency. In particular cases such as in forestry and tree planting, an agency such as National Forestry and Tree Planting Authority may serve as the required lead agency. The functions of the later include cooperating and coordinating with the National Environment Management Authority and other lead agencies in the management of forest resources.[73] The Authority also has the function of working in conjunction with other regulatory authorities to control and monitor industrial and mining developments in central forest reserves.[74] Where sand is to be mined for industrial and/or commercial purposes in forest reserves, the Authority will, on the basis of these statutory functions, have a say. The Authority will coordinate with the body created under the Mining Act for mining activities. The Mining Act provides for the office of Commissioner for Geological Survey and Mines Department to be appointed by the president.[75] The department is under the Ministry of Mineral and Energy Resources. The Commissioner has the powers to enter any land for the purpose of inspecting mining operations to ascertain whether the operations are in compliance with the law.[76] The Commissioner also has the power to take soil samples for examination purpose.[77]

8   Lessons from Other Jurisdictions

Under Kenya mining regime, minerals are described as a geological substance whether in solid, liquid or gaseous which occurs naturally in or on the earth, in or under water and in mine waste.[78] It includes the minerals specified in the First Schedule but does not include petroleum, hydrocarbon gases or groundwater.[79] This

---

[72]See Sec. 30(1)(a)(b) NEA.

[73]Sec. 54 (f) National Forestry and Tree Planting Act (2003).

[74]Sec. 54(g) National Forestry Act.

[75]Sec. 13(1) Mining Act.

[76]See Sec. 14(1)(a) b) Mining Act.

[77]Sec. 14(1)(c) Mining Act.

[78]Sec. 4 Kenya Mining Act (2016).

[79]Sec. 4 Kenya Mining Act.

is similar to the definition under the Ghana Mining Act 2006 which describes "mineral" as a substance in solid or liquid form that occurs naturally in or on the earth, or on or under the seabed, formed by or subject to geological process including industrial minerals but does not include petroleum as defined in the Petroleum (Exploration and Production) Law, 1984 (P.N.D.C.L. 84) or water.[80] Apart from the fact that sand can easily be accommodated under those generic descriptions which only specifically excludes petroleum, hydrocarbon gases and groundwater, the Kenya law also particularly provides for a category of minerals christened "construction minerals." In terms of the law, "construction minerals" includes stones, gravel, sands, soils, clay, volcanic ash, volcanic cinder and any other minerals used for the construction of buildings, roads, dams, aerodromes and landscaping or similar works."[81] It also includes such other minerals that the Cabinet Secretary declare, by gazette, to be construction minerals.[82]

For construction and other purposes, Kenyan National Sand Harvesting Guidelines 2007 is a statutory instrument which governs sand mining. A synonymous word, sand harvesting, is used for the process and it is defined as "the removal, extraction, harvesting or scooping of sand from designated sites."[83] The law provides for on-farm sand harvesting and Lakeshore/Seashore Sand Harvesting. It provides that on–Farm Sand harvesting and Lakeshore/Seashore Sand Harvesting should not exceed six (6) feet in depth.[84] It also provides that harvesting should be done concurrently with restoration of areas previously harvested which restoration will be undertaken with guidance from the Technical Sand harvesting Committee.[85]

These particular regulations under the National Sand harvesting Guidelines provide particular protection for agricultural lands in additional to specific environmental guidelines regulating mining activities under Kenyan laws which provide a good example for Uganda. Under South African laws, sand-mining is provided for under the legal regime governing mineral activities. The Mineral and Petroleum Resources Development Act is the primary statute that governs and regulates mineral resources and the process of exploiting same. Sand-mining is governed by the Act. The Act vests all mineral resources in the Republic of South Africa. According to the Act, a person wishing to engage in sand mining must apply to the relevant government for the right to do so in addition to having to comply with the requirements arising from comprehensive regulatory framework governing the exploitation of a mineral resource in form of various rights, permissions and permits.[86]

---

[80]Sec. 1 Kenya Mining Act.

[81]Sec. 1 National Sand Harvesting Guidelines (2007).

[82]Sec. 1 National Sand Harvesting Guidelines.

[83]Sec. 2 National Sand Harvesting Guidelines.

[84]Sec. 7(a) National Sand Harvesting Guidelines.

[85]Sec. 7 (c) National Sand Harvesting Guidelines.

[86]Mineral and Petroleum Resources Development Act (1994).

According to the Act mineral "means any substance, whether in solid, liquid or gaseous form, occurring naturally in or on the earth or in or under water and which was formed by or subjected to a geological process, and includes sand, stone, rock, gravel, clay, soil and any mineral occurring in residue stockpiles or in residue deposits, but excludes- (a) water, other than water taken from land or sea for the extraction of any mineral from such water; (b) petroleum; or (c) peat."[87] Based on this definition, The South African Department of Mineral Resources classifies sand as a naturally occurring industrial mineral which is a mineral that is mined for the value of its non-metallic characteristics. One fundamental feature of the South African mining law is the direct reference to both sand and soil in the general definition of minerals. Apart from the applicability of provisions relating to minerals in general to sands, the Act also makes specific provisions for sand mining based on the peculiar nature of sand. For example an applicant for sand mining would not, *ceteris paribus*, need to seek any prospecting or reconnaissance permissions or rights because deposits of natural sand are pretty identifiable and therefore no prospecting is needed to find deposits of sand.

The law requires that if sand mining is carried out on a sand mine within a period of two years and is restricted within a perimeter of 1.5 ha or less in which case the miner will need to apply for a 'mining permit' to start mining. If the sand mining extends beyond 1.5 ha and is done for a period of more than two years then the miner should apply for a 'mining' right' under section 22 which would grant the person mining the right to mine the deposit for a period of up to thirty years.

One downside of this is that sand miners may exploit this kind of provision to restrict their mining area to the 1.5 ha perimeters and, after two years, move to another acreage of land elsewhere place or even neighboring area and legalise same by making an application for a mining permit for yet another 1.5 ha.

This allows the miners to continue to mine sand around the country without necessarily applying for a permit. This kind of provision should not be adopted in Uganda as it is open to abuse.

9 Conclusions and Recommendations

The research has examined the legal regime of sand mining in Uganda. The chapter posits that that sand in Uganda is classified under mineral resources though in practice it is not so regarded. This is well demonstrated by the fact that royalties mandated to be leveled for minerals under the law are not leveled on sand mining by the relevant authorities. There are no comprehensive guidelines governing sand mining in Uganda, apart from the fact that the current provisions on sand as a mineral resource in the Mining Act can be clearer than they are now. Hence, Uganda requires a comprehensive regulation or guidelines setting out specific provisions on

---

[87]Sec. 1 Mineral and Petroleum Resources Development Act.

sand mining. The guidelines canvassed for here need to include specific provisions on approval, specification of restrictions like those relating to allowed depth of excavation in specific cases such as on farmlands as seen in the case of Kenya laws above cited, requirement for environmental impact assessment as well as issues relating to restoration of mined sites. It also needs to include provisions relating to shifting of dependency on sand and incentives for use of more ecologically better alternatives. This is in view of the point made that while sand mining has got a number of advantages: providing an essential raw material for infrastructural development for example road and housing works, providing raw materials for industrial products such as glass-making, serving as a source of employment for sand miners among others, the other side needs to be considered. The advantages need to be contextualized against its negative effects on the environment including agricultural lands so as to provide a basis for a strategic approach for dealing with it for example through a more robust regulatory framework to that effect.

References

AllAfrica (February 2nd, 2018) Uganda: Nema, Chinese firm lock horns over sand mining in L. Victoria. Available at: https://allafrica.com/stories/201802020179.html

Awaaz Foundation, Bombay Natural History Society (2013) Application for inclusion of sand mining in the Agenda of the Convention of Biodiversity, a new and emerging issue relating to the conservation and sustainable use of biodiversity. Available at: https://www.cbd.int/doc/emerging-issues/emergingissue-2013-10-Awaaz-Foundation-Bombay-NHS-en.docx

Daily Monitor (September 30th, 2016) Sand mining in Lwera: the dark side and bad deals. Retrieved from http://www.monitor.co.ug/artsculture/Reviews/Sand-mining-in-Lwera%2D%2DThe-dark-side-and-bad-deals/691232-3399554-gmfb9qz/index.html

Government of the Republic of Uganda, Ministry of Animal Industry, Agriculture and Fisheries (2013) National Agriculture Policy, NAP. Available at: http://agriculture.go.ug/wp-content/uploads/2019/04/National-Agriculture-Policy-1.pdf

Government of the Republic of Uganda, Ministry of Animal Industry, Agriculture and Fisheries, MAAF (2019) Priorities in sustainable soil management in Uganda. A presentation at the African Soil Partnership by Zakayo Muyaka, Assistant Commissioner, Soil & Water Conservation. Available at: http://www.fao.org/fileadmin/user_upload/GSP/docs/elmina/Uganda_Priorities.pdf

Government of the Republic of Uganda, Ministry of Land, Housing and Urban Development (2013) The Uganda Land Policy. Available at: http://extwprlegs1.fao.org/docs/pdf/uga163420.pdf

Government of the Republic of Uganda, National Environmental Authority, NEMA (2016) Uganda State of Environment Report 2016. The National Environment Management Authority Library (NEMAL), Kampala, Uganda

MPR News (March 8th, 2012) MPR News Primer: Frac sand mining. Available at: https://www.mprnews.org/story/2012/03/08/frac-sand-mining-mpr-news-primer

Naples Daily News (January 13th, 2017) Shrinking Shores: Florida sand shortage leaves beaches in lurch. Available at: https://eu.naplesnews.com/story/news/special-reports/2016/11/17/shrinking-shores-florida-sand-shortage-leaves-beaches-lurch/92052152/

Prabakhar VK (ed) (2000) Environmental pollution and awareness in the 21st century. Anmol Publications Pvt. Limited, New Delhi

Rajan G, Rao ASR (2007) Basic and applied soil mechanics, 2nd edn. New Age International Publishers

Saviour MN (2012) An environmental impact of sand and soil mining: a review. Int J Sci Environ Technol 1(3):125–134

The Guardian (February 27th, 2017) Sand mining: the global environmental crisis you've probably never heard of. Retrieved from https://www.theguardian.com/cities/2017/feb/27/sand-mining-global-environmental-crisis-never-heard

Trading Economics (2019) Uganda - agriculture, value added (% of GDP). Available at: https://tradingeconomics.com/uganda/agriculture-value-added-percent-of-gdp-wb-data.html

Wisconsin Department of Natural Resources (2017) Industrial sand mining in Wisconsin. A strategic analysis. Available at: https://dnr.wi.gov/topic/EIA/ISMSA.html

Overview of Main Challenges with Regards to Land Tenure in Africa: Factual and Legal Aspects

Mabel Munyuki-Hungwe and Mandivamba Rukuni

1   Introduction

Smallholder farmers make up the majority of farmers in Africa, yet they face many challenges including low productivity, low competitiveness, inadequate investment in technology and on-farm infrastructure as well as environmental degradation and vulnerability to climatic changes. The concerns with respect to customary tenure systems includes the fact that this land is largely unregistered, and that the administration of land under the traditional systems appears to work against effort to improve productivity.

These observations and associated concerns culminate in the fear that tenure security is compromised under customary law and its traditional system of administration As a broad context, it is important to recognise that African governments are shifting emphasis from trying to replace "customary" with "modern" tenure systems—an agenda based on the assumption that customary land rights systems have no scope for securing land rights and environmental management. Governments, however, are increasingly recognising that land policies and laws must build on local practice. *Cotulla* sums up this experience by observing that customary tenure systems are increasingly evolving in response to demographic growth, urbanisation, livelihood diversification, and cultural change.[1] Moreover, while early attempts at land titling in Africa were often unsuccessful, factors such as new legislation, low-cost methods, and increasing demand for land have generated renewed interest.[2]

---

[1]Cotula (2007).

[2]Deininger et al. (2011).

M. Munyuki-Hungwe · M. Rukuni (✉)
Barefoot Education for Africa Trust (BEAT), Marlborough, Harare, Zimbabwe
e-mail: ndaka@beatafrica.org

© Springer Nature Switzerland AG 2020
H. Yahyah et al. (eds.), *Legal Instruments for Sustainable Soil Management in Africa*, International Yearbook of Soil Law and Policy,
https://doi.org/10.1007/978-3-030-36004-7_7

Clarke paints a picture of communities across Sub-Saharan Africa that are increasingly having an officially recognised role managing communal land and local natural resources.[3] Most importantly, this acknowledges the clear links between land tenure and how people relate to their environment. These and other scholars do recognise the need for measured and wise interventions by governments, so as to improve governance capabilities of local institutions, rather than substitute them.

2   Framing Customary Land Tenure in Context of Customary Law

Because of the dominance of customary tenure in Africa, it becomes important to locate customary tenure as part or sub-set of a customary law systems, which in turn are part of traditional civilization, and belief/value systems or 'world-views'. International experience shows the challenge in reforming customary tenure on the basis of given (imposed) law, rather than reforming it in its own right and own evolutionary path.[4] In circumstances were given law drives reforms in customary law, the envisaged gains may not materialise easily, and/or new challenges and conflicts arise which also do not go away easily through a greater burden of statutory or given laws and regulation. Some scholars are now arguing that until African countries adopt legal pluralism, this challenge will not go away easily.[5] If the idea of legal pluralism is to be upheld, then there ought *"to be one legal system with two coequal sets of legal rules—received law and customary law—and the judicial system is empowered to fuse the systems over the long term. This equality means that communal and collective rights in land are recognized and protected, and people can choose one equal tenure and legal system over another.*

There is also local-level land administration and registration, where all customary interests are recorded and protected in land adjudication and customary as well as statutory alternative dispute resolution processes can be used. In addition, pluralist tenure and land law extend to urban areas, along with land regularization schemes and urban land adjudication".[6]

It is then feasible for the legal systems to draw from national goals and regulate so that legal rules are adapted to meet constitutional and relevant international norms relating to such priorities as gender equality, administrative justice, and protection of private and communal property rights. Experiences with customary tenure reforms show that African countries are increasingly adopting a pluralist approach. Africans more and more try to cease attempts to abolish customary tenure—given that abolition is an approach that is akin to 'running way from one's shadow'. Customary

---

[3]Clarke (2009).

[4]See Toulmin and Quan (2000) for further reference.

[5]See McAuslan (2005).

[6]McAuslan (2005).

law and beliefs that drive them never go away, and eventually governments are engaging rather than trying to abolish customary tenure. Moreover, customary tenure also has some strength that governments need to understand better and harness in the reform process.

The literature is also inconclusive when it comes to attributing major economic recovery of smallholder agriculture mainly to a land tenure reform.[7] This is increasingly seen as is rather unrealistic, given that secure land tenure will not resolve legacy issues of economic exclusion, poor infrastructure and remoteness from main markets, and the general inadequacy of public sector investment into these poor communities.[8] It can also be argued that absence of formal registration does not automatically mean insecure tenure. Moreover, the extent to which envisaged legal reform will unlock financial and economic value of land is also an issue that does not have much evidence of success in the international experience with customary tenure.[9] It would appear that most attempts to reform aim for the homogenization of national land laws so as to align all land with mortgage law. Customary land rights—however well protected and secured—don't appear to ever fulfill the requirements of commercial banks.[10] It can be argued therefore that it is conservative state bureaucracy and private mortgage practices and attitudes that need fundamental reform rather than customary tenure.

There is gravitation towards improving financial viability of rural businesses as the main means of improving competitiveness and bankability.

It would appear that where regulation is needed, the issue is what type of regulation as follows[11]:

- Government should avoid outdated colonially derived system that turn communities into squatters on their own lands, otherwise 'no regulation' may be better option;
- A critical role for the modern state is to establish ways of recognising and incorporating customary land tenure into national law, so that it is adaptable, dynamic and legitimate.
- Communities and civil society are increasingly recognising that a legitimate solution lies in protecting and strengthening customary land tenure and community governance.
- Government should therefore ensure that customary land rights are recognised as property rights in statutory law, not just as user or occupation rights, and have an equivalent force of law to private deeded property rights. Customary land must include not only the land of the family, house and farm but also forest, rangeland

---

[7]A discussion revived by the work of de Soto (2000).

[8]See i.a. Cousins (2002), Nyamu-Musembi (2006), Sjastaat and Cousins (2009) and Broomley (2009).

[9]Cousins (2002).

[10]Broomley (2009).

[11]See African Community Rights Network (2011).

and other lands held collectively, including those that are currently considered as state-owned.

- Government should recognise enormous benefits of recognising customary land tenure systems, and the huge problems created by giving primacy to foreign-derived law.

The best-case scenarios would start with legal pluralism as the long-term answer, with customary law and given/imposed law equal and the judiciary able to interpret both and fuse these as a long-term strategy.[12] This can lead into local-level land administration and registration, where all customary interests are recorded and protected in land adjudication and customary as well as statutory alternative dispute resolution processes can be used. Pluralist tenure and land law would extend to all areas—rural areas, rural towns and municipalities and urban areas. Participatory community planning replaces top-down master planning. Formal market institutions would come on board, and officials are advisers and facilitators of lay people who all are part of the decision-making process. There are clear rules for actions and trans-actions, as well as mechanisms for enforcement. That is mortgage law adapts to customary tenure.

The spirit of empowering rural people is noble. The devil is in the HOW. Governments need to learn from international good practice and some of it is covered below as follows[13]:

- The starting point is the statutory law recognising customary tenure rights **before** registration process. Registration should protect existing unregistered rights rather than undermine them (Uganda).
- Pilots used as a first step to implementation, with a strong monitoring and learning component to feed into subsequent planning and role out (Uganda, Mozambique).
- Creation of an outside boundary recorded/registered in the Surveyor General, with the rights on the inside held locally/by local office protects people from land theft by the State or investors. Information on the land rights of investors, state and the poor must be kept on the same record (Mozambique, Uganda, Namibia, South Africa).
- Registration is more effective when phased generally as follows:

    - Establish a decentralized system of land administration and adjudication down to community level.

      Demarcate and register the land administration zones and formalize land governance structures for each zone

    - Commence voluntary registration based on need.

---

[12]See McAuslan (2005).
[13]See Augustinus (2003).

- Finally go for systematic registration of remaining parcels as demand grows and the admin/adjudication system mature
- For example, in Uganda first phase focused on the production of a foundation for the purposes of land management and spatial information management at scale, also to address the land management issues of the day. The second phase used this foundation to increase the tenure security of households and individuals. Those people who have been demarcated who want titles can apply for customary certificates. This process is cheaper because it is systematic, and it deals with all disputes in an area at one time, thereby protecting the rights of the poor.
- While systematic titling is not a Best Practice, Best Practice involves the systematic collection and display of spatial and land record information for all decision makers, and the systematic cleaning up of titles and solving of ambiguities/disputes in preparation for individual titling on demand (Uganda; KwaZulu-Natal, South Africa).
- Democratic structures forming part of new land laws, which are meant to make land administration transparent, solve conflicts over land and bring good governance at the local level, can be too costly to implement in terms of the institutional structure required (Uganda, South Africa).
- Registry systems have to be drastically adapted to be able to register customary type tenures, not just by decentralization and transparent procedures and a user-friendly culture. Procedures have to be created which can move uncertain customary information about inheritance etc. to certain information on the registry record (Namibia).
- Decentralized implementation requires inter-agency co-ordination between lands, local government, justice etc. (Uganda, Mozambique)
- Conventional registries do not assist the public with legal advice. Registries for the poor need to also assist the poor with legal advice about their land rights and their options (Rehoboth-Namibia).
- New laws making provision for registration of group rights (Uganda, Mozambique, South Africa). Uganda has introduced the customary certification of these rights. Also, it is not individual certification, and a number of people and a range of rights can be registered such as the wife and husband and children, as well as those people with third party rights, such as people crossing the land, obtaining firewood etc. (Uganda).
- Protection of spouses through co-ownership laws (Mozambique) and joint estates through family law (South Africa) and tenure security through requiring consent for the transfer of land (Uganda, Kenya, South Africa).

3 International Experience with Communal Tenure

In recent years, African governments have shifted emphasis from trying to replace "customary" with "modern" tenure systems, and recognising that land policies and laws must build on local practice. *Cotulla* sums up this experience with reforms in customary tenure as diverse combinations of "statutory" and "customary" entitlements, and multiple and overlapping rights over the same resource. Customary tenure systems are increasingly evolving in response to demographic growth, urbanisation, monetarisation of the economy, livelihood diversification, greater integration in the global economy, and cultural change.[14] Customary tenure draws legitimacy from "tradition" yet these have been profoundly changed by decades of colonial and post-independence government interference, and are continually adapted and reinterpreted as a result of social, economic, political and cultural change. Local context largely determines dynamism and degree of change. In some instances, traditional authorities have maintained or increased their power and for others the powers have been severely eroded.

Often the traditional authorities have to enter various strategic alliances—with central government, local government bodies, political and business elites, and so on.

Moreover, while early attempts at land titling in Africa were often unsuccessful, factors such as new legislation, low-cost methods, and increasing demand for land have generated renewed interest.[15] In his celebrated work "Securing Communal Land Rights to Achieve Sustainable Development in Sub-Saharan Africa: Critical Analysis and Policy Implications", *Clarke* paints a picture of communities across Sub-Saharan Africa that are increasingly having an officially recognised role managing communal land and local natural resources.[16] This positive shift allows the resources that are vital for rural livelihoods and people's way of life to be managed by those that depend on them.

Most importantly, it acknowledges the clear links between land tenure and how people relate to their environment. Studies across the developing world demonstrate how environmental degradation worsens where tenure is unclear or not upheld.[17] Securing communal rights of access and usage is therefore crucial to the effectiveness of any scheme which empowers communities to manage communal land, especially with respect to sustainable management of the 'commons', that is—areas that comprise common pool resources, including land, water and forests.

---

[14]Cotula (2007).

[15]Deininger et al. (2011).

[16]Clarke (2009).

[17]See for example the findings in Clover and Eriksen (2009) for Botswana, Mozambique, South Africa and Zimbabwe.

3.1   Need for 'Tailored' Government Intervention

Cotulla goes on to argue that there is need for 'tailored' government action that builds on local practice, even where customary systems seem to work well at the local level.[18] This has become important lately as powerful outsiders such as urban elites and foreign investors who do not feel bound by those systems are exerting demand pressure, and government intervention may be required in these circumstances. In these cases, lack of legal protection for local land rights based on customary systems may result in local resource users losing land access. And, whether customary systems are still working well or not, *Cotulla* continues his argument, government intervention may be necessary to secure the resource claims of weaker and more vulnerable groups.[19] Government intervention in customary systems often also comes with attempts to further reform the inequitable aspects of customary tenure such as social status, age, gender and other aspects. Poor conceptualization and poor implementation of reforms, however, often lead to 'capture' by local elites (political, business, and NGOs) who then steer that change to suit their interests, while weaker groups lose out.

3.2   De jure and de facto Rights

One of the challenges in formalized reforms of communal tenure is failure to acknowledge the important distinction and dichotomy between *de facto* and *de jure* rights to the commons.[20] While according to formal legal instruments the state generally has primary ownership of the commons across Sub-Saharan Africa, de facto rights originate from users and generally do not receive state recognition. Although de facto rights are contained within a group-based system, individual access rights generally co-exist within the broader communal framework. Importantly, for the majority of Sub-Saharan African communities, these de facto rights are grounded in customary norms, enforced by indigenous legal systems and may receive state recognition only to the extent customary law is recognised. Any analysis of tenure arrangements over the commons therefore requires engagement with customary, indigenous law or non-state law. Customary law has been the subject of increasing scholarship in recent decades.[21] It can be defined as legal rules and processes that have become an intrinsic part of accepted legal conduct and arise from social practices rather than positive law.[22] For Sub-Saharan contexts it is

---

[18]Cotula (2007).

[19]Cotula (2007).

[20]Clarke (2009).

[21]See i.a. Bennett (2004), Cotula (2007) and Tobin (2013).

[22]For a more extensive analysis of the nature of customary law in the African context see Elias (1956).

usually based on indigenous norms and value systems. Today there is now widespread recognition that customary principles and dispute resolution processes represent the most relevant legal framework for rural communities in Sub-Saharan Africa. Current literature emphasises the dynamic, adaptive and flexible capacity of customary legal systems, dispelling widely held conceptions regarding their static, 'traditional' nature.[23]

These authors also provide a further distinction between 'communal tenure' and 'customary law'.[24] Customary law governs tenure over the commons in the majority of rural Africa. Communal tenure can be defined as—community rights to common property and resources—and customary law as—the indigenous norms and institutions that govern access and use. Despite a trend towards increasing formalisation of land tenure, particularly in urban and peri-urban areas, customary land management institutions are still the most important authority regulating communal land administration. Regardless of this reality, until recently customary authorities have been severely marginalised in land reform policies, primarily due to the centrist nature of reform and the presumed insecurity of tenure they afford.

In a positive shift, there is now greater recognition that communal tenure and customary land institutions must be an integral part of land administration. Improving security of tenure in Africa has remained a controversial, politically charged and potentially destabilising issue.

This is occasioned by the colonial legacy of legal pluralism, mass expropriation of land, marginalisation of indigenous tenure systems, all leading to uncertainty over land tenure.

Large-scale national reforms have been widely criticized in the literature.[25] In most cases effective implementation has proved unachievable or has actually resulted in increased tenure insecurity. Even after decades of reform towards formalisation, it has been estimated that only 2–10 per cent of rural land across the continent has been formally titled.[26] Numerous studies have demonstrated the negative consequences that titling processes can cause. *Meizen-Dick et al.* demonstrate the risks inherent in titling policies, particularly for secondary rights users, due to elite capture and increasing inequality in land ownership.[27] *Nyamu-Mesembi* further highlights the failure of formal title to increase access to credit in rural Africa, despite assumptions to the contrary.[28] As a result, current literature emphasises reforms that build on existing indigenous practices, empower local communities, avoid one-size- fits-all solutions, and appreciate the political and social dynamics

---

[23]Schlager and Ostrom (1992), Ouedraogo and Toulmin (1999) and Cotula (2007).

[24]Schlager and Ostrom (1992), Ouedraogo and Toulmin (1999) and Cotula (2007).

[25]See i.a. Walsh (1993), Place and Migot-Adholla (1998) and Okoth-Ogendo (2014).

[26]Clarke (2009).

[27]Meizen-Dick et al. (2008).

[28]Nyamu-Musembi (2006).

inherent in land relations.[29] Formalising tenure in law is but one issue among many in the rural socio-political process, and by no means a sufficient condition.

Clarke concludes further that post-independence government policies have favoured a centralised, top-down approach in tenure reforms, assuming it is the most effective in ensuring productive exploitation.[30] Lack of and inadequate state capacity and enforcement, however, made centralised efforts to manage the commons ineffective and generally led to de facto open access regimes.[31]

Since the 1980s there has been increasing recognition that despite statutory reforms to limit their authority, in many cases customary authorities remain effective at managing access to communal land and regulating sustainable exploitation.[32]

This growing interest in customary management systems, combined with governance trends towards decentralization, citizen participation and the prominence of sustainable development discourse, have laid the foundation for an officially recognised role for communities in managing the commons.

Policy regarding community management of the commons has in recent times been dominated by the community-based natural resource management (CBNRM) approach.[33] While primarily a technical (as opposed to legal) methodology, CBNRM recognises the importance of secure rights of access and exclusion as a basic precondition of effective management. further requires a well-defined social group, clear regulations that limit levels of exploitation and a capacity to monitor and enforce the rules. Rather than a prescriptive policy, CBNRM is more a model that requires modification due to the nature of the particular resource, the technical capability of government authorities, local power dynamics and external interests regarding extraction. Despite the need for context specificity, CBNRM is today widely considered the most effective strategy to balance the needs of sustainable exploitation and environmental management.[34] Current policy regarding tenure of the commons appears to have learnt the mistakes of centralised titling programs.

To dismiss customary institutions as unrepresentative and undemocratic fails to appreciate the reality many rural communities face. Across the region, large-scale attempts at land reform have had minimal impact and the state rarely has a legitimate presence in rural areas regarding land affairs. So if it is accepted that customary institutions are generally best placed to manage communal land and resources, then the question remains how can equity and accountability be achieved in undemocratic institutions? Quota systems can be put in place to guarantee representation of certain vulnerable groups such as women and minorities, however this may not address how

---

[29]See i.a. McAuslan (2005), Cotula and Mathieu (2008), Meizen-Dick et al. (2008) and Monteiro et al. (2014).

[30]Clarke (2009).

[31]Hilhorst and Aarnink (1999).

[32]See i.a. Sheperd (1992), Chimhowu and Woodhouse (2006) and Biitir and Nara (2016).

[33]See for example the CBNRM approach in Ngamiland, Botswana examined in Kgathi and Ngwenya (2005).

[34]See the analysis provided in Dressler et al. (2010) for further reference.

decisions are made. More appropriate may be establishing effective oversight measures and linking community-level groups into the broader regulatory framework.

4 Conclusion

Evidence on Africa's experience with communal tenure reforms is that the true 'tragedy of the commons' is that despite their vital role in rural development, these common resources are often trapped between dysfunctional land legislation that vests ownership in the state and unrecognised communal tenure administered by customary authorities.

Efforts to clarify tenure arrangements and improve security of tenure, while grounded in a sound theoretical approach, have failed to achieve the intended impact.

This situation has led to de facto open access regimes that allow unsustainable resource depletion and land degradation. Given the pressures of projected population growth, increased resource demand and a trend towards privatisation of communal land, the commons are under increasing threat. Unclear and ineffective tenure arrangements only exacerbate the situation. Practical solutions are therefore needed now more than ever. If the implementation issues can be overcome, increasing security of communal tenure can provide a basis for more sustainable management of the commons and offers hope that the sustainable development promised under international law can be more than just rhetorical.

References

African Community Rights Network (2011) Statement to Governments from the African Community Rights Network Douala Conference on Community Rights Cameroon, 13–16 September 2011. Available at: https://www.forestpeoples.org/en/topics/rights-land-natural-resources/publi cation/2011/statement-governments-african-community-rights

Augustinus C (2003) Comparative analysis of land administration systems: African review with special reference to Mozambique, Uganda, Namibia, Ghana, South Africa. Work Undertaken for the World Bank, Funded By DFID January, 2003

Bennett TW (2004) Customary law in South Africa. Juta and Company Ltd., Landsdowne

Biitir SB, Nara BB (2016) The role of Customary Land Secretariats in promoting good local land governance in Ghana. Land Use Policy 50:528–536. Available at: http://www.sciencedirect. com/science/article/pii/S026483771500335X

Broomley DW (2009) Formalising property relations in the developing world: the wrong prescription for the wrong malady. Land Use Policy 26(1):20–27

Chimhowu A, Woodhouse P (2006) Customary vs private property rights? Dynamics and trajectories of vernacular land markets in sub-Saharan Africa. J Agrar Change 6:346–371

Clarke RA (2009) Securing communal land rights to achieve sustainable development in sub-Saharan Africa: critical analysis and policy implications. Law Environ Dev J 5(2):130

Clover J, Eriksen S (2009) The effects of land tenure change on sustainability: human security and environmental change in southern African savannas. Environ Sci Policy 12(1):53–70

Cotula L (ed) (2007) Changes in 'customary' land tenure systems in Africa. International Institute for Environment and Development, London

Cotula L, Mathieu P (2008) Legal empowerment in practice: using legal tools to secure land rights in Africa. Available at: http://pubs.iied.org/12552IIED.html

Cousins B (2002) Reforming communal land tenure in South Africa – why land titling is not the answer: critical comments on the Communal Land Rights Bill 2002. Program for Land and Agrarian Studies (PLAAS), University of Western Cape, South Africa

de Soto H (2000) The mystery of capital: why capitalism triumphs in the West and fails everywhere else. Basic Books, New York

Deininger K, Ayalew Ali D et al (2011) Ethiopia: rural land certification in Ethiopia: process, initial impact, and implications for other African countries. World Dev 36(10):1786–1812

Dressler W, Büscher B, Schoon M et al (2010) From hope to crisis and back again? A critical history of the global CBNRM narrative. Environ Conserv 37(1):5–15

Elias TO (1956) The nature of African customary law. Manchester University Press

Hilhorst T, Aarnink N (1999) Co-managing the commons: setting the stage in Mali and Zambia. Royal Tropical Institute

Kgathi D, Ngwenya B (2005) Community based natural resource management and social sustainability in Ngamiland, Botswana. Botswana Notes Records 37:61–79

McAuslan P (2005) Legal pluralism as a policy option: is it desirable? Is it doable? Collective Active for Property Rights (CAPRi) Policy Briefs. From the proceedings of the Workshop *"Land Rights for African Development: From Knowledge to Action."* Hosted from October 31st to November 3rd, 2005 by the UNDPs Drylands Development Center and the International Land Coalition

Meizen-Dick R, Di Gregorio M, Dohrn S (2008) Decentralization, pro-poor land policies, and democratic governance. CGIAR Systemwide Program on Collective Action and Property Rights (CAPRi), Working Paper No. 80, Washington D.C.

Monteiro J, Salomão A, Quan J (2014) Improving land administration in Mozambique: a participatory approach to improve monitoring and supervision of land use rights through community land delimitation. Paper prepared for presentation at the 2014 World Bank Conference on Land and Poverty. The World Bank - Washington DC, March 24–27, 2014

Nyamu-Musembi C (2006) Breathing life into dead theories about property rights: de Soto and land relations in rural Africa. Institute of Development Studies, Working Paper 272

Okoth-Ogendo (2014) Legislative approaches to customary tenure and tenure reform in East Africa

Ouedraogo H, Toulmin C (1999) Land tenure, poverty and sustainable development in West Africa: a regional overview. International Institute for Environment and Development

Place F, Migot-Adholla SE (1998) The economic effects of land registration on smallholder farms in Kenya: evidence from Nyeri and Kakamega districts. Land Econ 74(3):360–373

Schlager E, Ostrom E (1992) Property-rights regimes and natural resources: a conceptual analysis. Land Econ 68(3):249–262

Sheperd G (1992) Managing Africa's tropical dry forests: a review of indigenous methods. Overseas Development Institute, London

Sjastaat E, Cousins B (2009) Formalisation of land rights in the South: an overview. Land Use Policy 26(1):1–9

Tobin BM (2013) Bridging the Nagoya compliance gap: the fundamental role of customary law in protection of indigenous peoples. Resource and knowledge rights. Law Environ Dev J 9(2):142

Toulmin C, Quan J (eds) (2000) Evolving land rights, policy and tenure in Africa. DFID/IIED/NRI, London, UK

Walsh MT (1993) The social and economic impacts of land reform: a Kenyan case study. Paper presented to the East African Seminar Series, African Studies Centre, University of Cambridge, 9 November 1993

African Feminism, Land Tenure and Soil Rights in Africa: A Case of Uganda

Godard Busingye

1 Introduction

The African continent has various forms of land tenure systems, each tenure having unique characteristics and bestowing specific land and soil rights to the holder. Land tenure systems in Africa are a creature of western colonial powers and pose serious problems to land and soil rights of Africans. This chapter uses the lenses of African feminism, particularly the motherism brand to unpack, repackage and recommend reconstruction of land tenure systems to benefit all Africans without discrimination. In Africa, land rights are closely related to soil rights, with no clearly marked differences. For example, in Uganda there is no difference between the concept of land and soil.

Among the Baganda in the central region, the word '*etaaka*' refers to the land and also the soil; in the western part of the country, the *Runyakitara* speakers use the word '*eitaka*' to refer to land and soil as well.[1] The longevity of land rights under particular tenure systems is also contemporaneously equivalent to that of soil rights. Right to acquire land rights under a particular tenure system is, in majority of cases, determined by right of citizenship. Citizenship may be acquired at birth but may also be acquired or even lost at much later time. For purposes of this chapter, land tenure refers to the value of interest a person has in a particular parcel of land while soil rights refer to the bundle of rights a person gets over land, including the right to utilize the land and staying on it either as a citizen or alien. In essence, the right to soil is a theoretical right attached to citizenship, the latter being the determinant

---

[1] *Runyakitara* language unites the people of the western region of Uganda right from the Lake Albert and River Nile moving southwards up to the borders with the Republic of Rwanda. Ethnicities in this region include the Banyoro, Batoro, Bakonjo, Bamba, Banyankore and Bakiga.

G. Busingye (✉)
Kampala International University, School of Law, Kampala, Uganda

© Springer Nature Switzerland AG 2020
H. Yahyah et al. (eds.), *Legal Instruments for Sustainable Soil Management in Africa*, International Yearbook of Soil Law and Policy,
https://doi.org/10.1007/978-3-030-36004-7_8

factor in bestowing entrenched or limited land or soil rights to the holder. The chapter uses Uganda as a case to demonstrate the relationship between African feminism, land tenure and soil rights in Africa. The case study method enables the analysis made to identify the problematic misconceptions about land and soil rights as constrained by the various land tenurial systems identified in the chapter. Paradoxically, human life cannot be sustained on land without the land right holder enjoying soil rights. The relationship between land rights of holders and soil rights, therefore, is intrinsic and mutually reinforcing. It can be clearly understood using the lenses of African feminism perspective, which explains, albeit in a theoretical manner, the socio-legal foundations of each of these rights.

Rights to soil basically fall into two categories; *jus soli* (automatic soil rights based on birth right citizenship) and *jus sanguinius* (soil rights based on right of blood or familial lineage). *Jus sanguinius* is the prevalent form of soil rights in most African countries.[2] The discussion in this chapter, therefore, largely hinges on the *jus sanguinius* rights, which are tagged to blood or familial lineage a person has at the time of birth. For instance, children born in Uganda, but whose parents are not citizens of Uganda, or whose parents are not known to be Ugandan citizens do not get automatic citizenship and soil rights at birth. They, however, may be presumed to be citizens of Uganda by birth if up to the age of five, their parents remain unknown.[3] Land tenure systems in Uganda are of two broad categories, those that grant perpetual interest in land, and hence accord perpetual soil interests to the land holder and those that grant only limited interest in land, and equally, limited access rights to the soil.

Citizens of Uganda are entitled to acquire perpetual interest in land and also in soil under customary, freehold or *mailo* tenure systems; non-citizens, can only acquire limited rights to soil, under the leasehold tenure system.[4]

The main argument in this chapter is that acquisition of land and soil rights are important because they bestow firm rights to each of them and entice the rights holders to sustainably utilize the same. It therefore, becomes important to use a theoretical framework such as the African feminism to highlight the socio-legal regime governing land and soil rights, which historically were constructed within the realm of the ideology of patriarchy, and intended largely favour men, and to a limited

---

[2]See the Constitution of Uganda, 1995, Article 10 which provides: [T]he following persons shall be citizens of Uganda by birth (a) every person born in Uganda one of whose parents or grandparents is or was a member of any of the indigenous communities existing and residing within the borders of Uganda as at the first day of February, 1926, and set out in the Third Schedule to this Constitution; and (b) every person born in or outside Uganda one of whose parents or grandparents was at the time of birth of that person a citizen of Uganda by birth.

[3]The Constitution of Uganda, 1995, Article 11(1) provides: A child of not more than five years of age found in Uganda, whose parents are not known, shall be presumed to be a citizen of Uganda by birth. Article 11(2) provides: A child under the age of eighteen years neither of whose parents is a citizen of Uganda, who is adopted by a citizen of Uganda shall, on application, be registered as a citizen of Uganda.

[4]Perpetual interests in land and soil accrue to anyone who is, or who becomes a citizen of Uganda, without discrimination.

extent, citizens. The patriarchal construction of society created a stratified gendered socio-legal arrangement where women in particular, citizens or non-citizens, are only expected to acquire land and soil rights through the forms of rights of their male acquaintances, husbands, brothers or fathers. Under this arrangement, the male acquaintances are in position to determine how the land and soil rights can be acquired and accessed and how each can sustainably be utilized by any family member.[5] The next section discusses how socio-legal stratification affects both land and soil rights in Africa, using the case of Uganda.

2 Socio-Legal Stratification, Soil Rights and Land Tenure Systems

The gendered nature of land and soil rights in Africa remains a problem in spite of the global steps taken towards bridging the gaps between women and men in the enjoyment of socio-legal and economic rights. The United Nations General Assembly recognized this fact while adopting the Beijing Declaration in 1995. During the Beijing Conference, the United Nations General Assembly *inter alia* stated:

> [T]he status of women has advanced in some important respects in the past decade, but that progress has been uneven, inequalities between women and men have persisted and major obstacles remain, with serious consequences for the well-being of all people.[6]

The obstacles alluded to in the Beijing Declaration include custom, colonial law, colonial history, colonial education, and religion all of which are informed by the ideology of patriarchy—the rule of the fathers.[7] These obstacles create a stratified and sectionalized environment where accessing land and soil rights within the existing land tenurial systems remain skewed and tilted in favour of the male gender and the citizens.[8] Within these stratified and sectionalized social structures, men and citizens are the perceived social and legal owners of the land and soil.[9]

Land tenure systems being the avenue through which land and soil rights can be accessed, determine how an individual may utilize a particular parcel of land and its soil. Access to land and soil rights in Uganda, therefore, cannot be divorced from the patriarchal ideological contestations that created the land tenurial systems. Land tenure systems in Uganda are provided for in the Constitution, 1995 and the Land Act, Cap. 227 as: customary, freehold, mailo and leasehold.

---

[5]Kagoda (2008).
[6]United Nations (1995).
[7]Government of Uganda, Land Policy (2013).
[8]Busingye (2012).
[9]Busingye (2017).

The ideology of patriarchy, which conditions land and soil rights in Africa has is its roots in the historical Greek philosophy. Historicizing patriarchy, Napikoski & Lewis aver:

> [A]patriarchy, from the ancient Greek patriarches, was a society where power was held by and passed down through the elder males. When modern historians and sociologists describe a 'patriarchal society', they mean that men hold the positions of power and have more privilege: head of the family unit, leaders of social groups, boss in the workplace, and heads of government.[10]

Much as the ideology of patriarchy is the foundation for most land tenure systems in Africa, African land tenure systems are varied, and may not necessarily have the same impact on women and men and citizens and non-citizen's land and soil rights.[11] The varied nature of land tenure systems in Africa, therefore, makes it unrealistic to talk of an African land tenure system that regulates access to soil rights.[12] Moreover, the typology of African soils, which also determines how soils are utilized, varies from region to region and country to country on the African continent. Land use patterns are equally diverse on the continent. Much as it may not be possible to generalize land tenure systems in Africa, and hence come up with a harmonized position about continental land and soil rights, land tenure systems on the continent can conveniently be categorized as private, communal, open access and State owned tenures.[13] Under the communal setting, access to soil rights, are constructed by, and specifically intended to serve the rights of the patriarch—male heads within the stratified socio-legal communal set up at the expense of women. The situation may not be so different under the private, open access and state owned land tenure systems. Each of these systems is either supported by custom or laws constructed and enacted within a male oriented patriarchal society. On its part, custom is society specific and excludes non-societal members, in the same way as citizenship excludes non-citizens from automatically accessing land and soil rights in a particular country.

The foregoing analysis makes it possible to understand that the norms embedded in custom and the law, which define how individual women and men, citizens and non-citizens acquire land and soil rights suffer from the inherent constraints in the ideology of patriarchy, upon which they are constructed and interpreted.[14] The patriarchal society is stratified and does not benefit every member of the socio-legal class equally, women and men.[15]

---

[10]Napikoski and Lewis (2018).

[11]For example, Chad, Lesotho and the United Republic of Tanzania, which have unrestricted *jus soli*, cannot be said to have the same problems associated with soil access rights like the majority of other African states which rely on the principle of *jus sanguinius*.

[12]FAO (2010).

[13]Wily (2011).

[14]Muinde (2013).

[15]Saunders (1990).

Within the confines of the ideology of patriarchy and necessarily the stratified and sectionalised society, however, men, whether citizens or not, have more chances to land and soil rights than women.[16] In this chapter, African feminism helps to identify the ideological tools of patriarchy, deconstruct and reconstructed them to enable women and men access equal land and soil rights.[17] Unpacking the socio-legal stratified society is intended to make it clear that land and soil rights and land tenurial systems in Africa are inter-linked and reinforce each other within the stratified and gendered society. Based on this analogy, the next subsection identifies, through an African feminism theoretical framework, how land tenure systems that regulate access to land and soil rights can be repackaged to benefit women and men and citizens and non-citizens without discrimination.

3   The Link Between African Feminism, Land Tenure Systems and Soil Rights in Africa

The foregoing discussion highlighted that the theoretical, factual and legal rights to land and soil rights are problematic and are constructs of the ideology of patriarchy. Those rights are enjoyed differently by women and men and citizens and non-citizens. This section uses the African feminism theoretical perspective to interrogate the ideological contestations in the land and soil rights regime in Africa, using the case of Uganda. African feminism in this particular case uses analogy of motherism (mother's love to her children) to drive the nail in the socio-legal fabric constructed by the ideology of patriarchy.[18] In so doing, the analysis unpacks and attempts to repackage the operational mechanisms in the existing land tenure systems in a better design capable of engendering sustainable utilization of the African land and soil. The analysis maintains a focus on the need to equitably redistribute land to women and men, citizens and non-citizens, without discrimination, in order to enable equitable access to land and soil rights.

According to the African feminism, land tenure systems in Africa lean towards bestowing more access to land and soil rights to men than to the women.[19] It suffices to note that African feminism, often dominated by debates on womanism (the biological factor of being a female, and also the weak gender), is a brand of the general feminism perspective and shares a lot with other brands of feminism.[20]

A deeper understanding of the African feminism, and necessarily African womanism perspective, reveals that it shares a lot with the western written forms of feminism. In this regard, Mwale asserts:

---

[16]Asiimwe (2001).
[17]Kandiyoti (1988).
[18]Mwale (2002).
[19]Asiimwe (2001).
[20]Coetzee (2017).

[A]frican womanism, despite its pretensions to seeking cooperation or its advocacy for interdependency between men and women, uses a model of conscientisation of women that is foreign to Africa, and runs the risks of obscurantism, vulgarism, inauthenticity, and irrelevance.[21]

African feminism is, however, embedded within the African custom and part and parcel of the African society and identifiable with the African person in his or her own right. It can, therefore, easily be distinguished from the Western forms of feminism by understanding how African men and women relate to land and utilize the soil. The relationship between an African person and the land is in many ways similar to his or her relationship with the soil because there is no significant difference between land and soil within the African domain. It is essentially the same relationship a mother has with her children, hence, the analogy of motherism in regard to the relationship between an African and land or soil. In the case of Uganda, the *Runyakitara* word, '*eitaka*', means either land or soil, the same applies to the *Luganda* word, '*etaaka*'. Contemporary African feminist attack the ideology of patriarchy and its *modus operandi* in terms of allocation of, or denial of land rights [and soil rights] to women and men, and citizens and non-citizens.[22] The radical feminism perspective encourages a very liberal usage of the term patriarchy to apply to virtually any form of male domination.[23] Socialist feminism is mainly restricted to analyzing the relationships between patriarchy and class under capitalism.[24] Under the capitalist modal of classification, femininity denotes the weak—the women, while masculinity denotes the strong and decision-makers—the men, hence the socio-legal stratum of society.[25]

In reality, African feminism attack of ideological of patriarchy conception goes well beyond the male/female divide. It also penetrates the feminine stratum which is equally classified it into the elderly, the mothers, the daughters, the in-laws, the illiterate and the elite and attempts to remodel it into a more accommodative structure.[26] In regard to land tenurial systems and hence land and soil rights in Uganda, sectionalization of society elevates all the male gender over all the female gender; and citizens over non-citizens. It equally elevates the sectionalized patriarchal families of the fathers over those over the mothers.[27]

In that regard, it creates a hierarchy of the landed, and the landless, the latter having limited access to soil rights.[28] Limited access to soil rights renders the whole concept of sustainable development untenable within the patriarchal African setting. Owners of the land and hence entrenched soil rights are not necessarily the ones that

---

[21]Mwale (2002).

[22]Chopra and Mülle (2016).

[23]Kandiyoti (1988).

[24]Ibid.

[25]Valledor-Lukey (2012).

[26]Busingye (2017).

[27]Busingye (2017).

[28]Gingrich (1999).

utilize the land, those that utilize the land, especially for agriculture purposes, have little stake in the land. The latter may over-exploit the soil without taking into account the need to conserve it, because they are not its recognized owners.

The uniqueness of African feminism in the drive to deconstruct and reconstruct land and soil rights lies in the fact that it is largely built on indigenous models of African motherism (love), African womanism (Queen mother), African femalism (greater respect for the female body), and snail-sense feminism (slow, but sure—women's adaptation strategy to patriarchal systems).[29] The implications of this theoretical analysis is that much as women and non-citizens are denied automatic access to land and hence soil rights by the patriarchy society, they nevertheless remain a potential force that can be relied upon to sustainably utilize the soil, if given a legal authority to do so. It is a fact that today; many industries and plantation farms in Uganda are owned by non-Ugandans, initially Indians of British origin, and now increasingly the Chinese.

African feminism paradigm being a medium through which the socio-legal phenomena that regulates land and soil rights is viewed and interpreted equally helps to critique the current move by contemporary African leaders to provide solutions to African specific problems, largely basing on harnessing the soil.[30] That is because the vision of sustainable development of African leaders is riddled with the western capitalist and exploitative paradigms that are not well integrated into the African social fabric.

The so proclaimed paradigm of African solution to African problems is largely financed by the Western donor community whose hidden interest is to suffocate the indigenous African womanism paradigm that loves and cares for the African land and soil.

That remains true much as paradigm of 'an African solution for African problems' appears to front the human rights regime reminiscent in the Western ideologies.[31] The latter cannot be accommodated within the African society without appropriate reconstruction measures capable of making it African use friendly. That thinking, however, provides a window of hope because it shows that Africans are aware that they are endowed with capacity to meritoriously reason and ably provide their own perspectives of the factual socio-legal phenomena regarding access to land and soil rights.

Together with the African feminism perspective, the African paradigm of 'African solution to African problems' may therefore, cautiously be relied upon to provide possible solutions to the inequalities in the romanticized land tenurial systems created by the western patriarchy ideology during the colonial era to regulate access to soil rights. In this respect, motherism brings on board dynamics of rebuilding society in

---

[29] Nkealah (2016) and Ezenwa-Ohaeto (2015).

[30] Mays (2003).

[31] Chirisa et al. (2014).

cooperation with mother nature at all levels of human endeavour.[32] There are chances that African feminism can create a conducive environment which in the long run may bring about sustainable utilization of African land and soils by women and men, citizens and non-citizens.

The basis for this reasoning is that under the African motherism perspective, women are equated to nature, which is perceived as caring, and motherly.[33] The motherism perspective is valuable because within its purview, nature cannot be degraded to a worthless component of human existence. Moreover, women constitute the largest percentage of persons that harness land in Africa.[34] As a matter of fact, most households in Africa and Uganda in particular, depend on the soil and are dependent on rain-fed agriculture as their main source of income.[35] The agricultural sector employs over three-quarters of the active labour force. Eighty three percent of Uganda's women are employed in the agriculture sector. Only seventy one percent of men are employed in the same sector.[36] Equal rights to land and soil rights, therefore, become a pre-requisite to sustainable development based on harnessing the soil.

In reality, the socio-legal stratification of society occurs because the ideology of patriarchy is a system of oppressive power relations that reorganizes society into sections of the privileged and under-privileged, based on privileged supremacy.[37] Sectionalism necessarily discriminates against the less privileged members of society in various aspects.[38] In such a problematic social order, the mothers, and non-citizens have fewer opportunities to access soil rights both under customary or the legal tenurial systems in their individual rights.[39] African feminism, has created a strong base of women agency that resists oppression of the ideology of patriarchy. Women, have, hence, through their agency, been able to gain a fair deal in accessing land, and hence access soil rights under the existing land tenure regime.[40]

Women's agency encompasses the revolutionary power of women to rebel against the existing socio-legal constructed land tenurial systems.[41] Non-citizens have equally been able to acquire long term leaseholds and large chunks of land and hence are able to exercise some degree of sustainable utilization of land and soil, because they are not threatened by imminent land evictions.

Through their agency informed by African feminism perspective, women in Uganda have for example influenced constitutional reforms. For example, Article

---

[32]Nkealah (2016).

[33]Walker (1995).

[34]Slavchevska et al. (2016).

[35]Asiimwe (2001).

[36]Okonya and Kroschel (2014).

[37]German (1981).

[38]Ahmed (1999).

[39]Busingye (2017).

[40]Namubiru-Mwaura (2014).

[41]Chopra and Mülle (2016).

33 (3) of the Constitution of Uganda provides: '[T]he State shall protect women and their rights, taking into account their unique status and natural maternal functions in society'.[42] With the women wriggling through their agency to transform society, stratification of society has now become archaic and dysfunctional. Its harmful societal values have equally been fundamentally weakened. In regard to non-citizens, and in a bid to grant them access to firm land and soil rights within the constraints of existing land tenurial systems in the country, the Constitution under Article 9 provides: 'Every person who, on the commencement of this Constitution, is a citizen of Uganda shall continue to be such a citizen'. The Constitution then provides for the modes of how a non-citizen of Uganda can acquire citizenship in the country.[43]

Once acquired, Ugandan citizenship may be lost, but cannot, thereafter be taken away arbitrarily because it becomes a human rights aspect of the holder.[44]

Indeed, Africa governments being part and parcel of the global community attempt to incorporate the philosophical foundations of the human rights regime envisaged under the Universal Declaration of Human Rights, 1948 into their national legal frameworks. Much as the human rights philosophical foundations provided for under the 1948 human rights framework are necessarily western ideals, they can be harmonized with the African feminism thinking to create a socially and legally balanced human rights framework for Africans. In this regard, the human rights approach in relation to access to land and soil rights under the existing land tenurial systems in Uganda fit well in the outfit provided by Gaard, who asserts:

> [t]he moral problem arises from conflicting responsibilities rather than from competing rights and requires for its resolution a mode of thinking that is contextual and narrative rather than formal and abstract. This conception of morality as concerned with the activity of care centres moral development around the understanding of responsibility and relationships, just as the conception of morality as fairness ties moral development to the understanding of rights and rule.[45]

This approach, therefore, helps to put into context the wraths of the ideology of western patriarchy, which, when viewed through the lenses of African feminism are clearly identified as the problematic arenas of oppression to women and

---

[42]Constitutional reforms in Uganda were contemporaneously adopted with the global agitation for women's equal rights in all spheres at the various United Nations Women Conferences, including the Beijing Conference in 1995.

[43]Article 12 of the Constitution provides for acquisition of Ugandan citizenship by registration, while Article 13 provides for acquisition of Ugandan citizenship by naturalization.

[44]Article 14 of the Constitution provides for how Ugandan citizenship may be lost, namely: A person may be deprived of his or her citizenship if acquired by registration, on any of the following grounds: (b) voluntary service in the armed forces or security forces of a country hostile to or at war with Uganda; (c) acquisition of Uganda citizenship by fraud, deceit, bribery, or having made intentional and deliberate false statements in his or her application for citizenship; and (d) espionage against Uganda.

[45]Gaard (1993).

non-citizens. Hartmann succinctly describes the fabric of western patriarchal ideology as:

> [A] set of social relations between men, which have a material base, and which, though hierarchical, establish or create interdependence or solidarity among men that enable them to dominate women. [T]he material base upon which patriarchy rests lies most fundamentally in men's control over women's labour power ... [It] does not rest solely on childbearing in the family, but on all the social structures which enable men to control women's labour. Control is maintained by denying women access to necessary economically productive resources.[46]

The analysis made in this section reveals that in the case of an African, soil rights are contemporaneously acquired with the acquisition of land rights under the various land tenurial systems on the continent. Land tenure systems are, however, constructed within the purview of the ideology of patriarchy, which discriminates against persons it designates as weak or foreigners, and deny them automatic soil access rights. Women and non-citizens are the victims of that ideology. The socio-legal foundations of land tenure systems on the African continent, therefore, need to be interrogated further using the lenses of African feminism for there to be a clear understanding of how problematic the situation is for those designated as weak and foreign within the purview of the ideology of patriarchy. Moreover, African feminism goes to the root of the African society and has been used by African women in particular to reinvigorate their agency that now resists African patriarchy manifestations in custom and the western forms of patriarchy that seek to exclude the women from accessing land and hence soil rights *in tandem* with the men. The discussion clearly brings out the symbiotic relationship between land tenure systems, soil rights and feminism perspective, the latter being the medium through which land tenure systems and soil rights in Africa are viewed, understood and interpreted.

The analysis made in this section, therefore, forms the basis for a discussion of the land policies and laws in the next subsection, which goes a step further in interrogating the role played by law and custom to deny certain persons automatic access to land and soil rights on the African continent.

4   Policy and Legal Regulation of Access to Soil Rights

4.1   Contextual Perspectives

The previous section discussed how African feminism theoretical framework can be relied upon to influence attitudinal change in access to land and soil rights on the African continent. African feminism specifically faulted both the African custom and the western forms of the ideology of patriarchy as being responsible for the woes of Africans in regard to land and soil rights. This section builds on the momentum

---

[46]Hartmann (1979).

gained in the previous section to critique policy and legal regulation of land and soil rights as embodiments of the oppressive ideology of patriarchy. The current policy and legal norm in majority of African countries was exported to Africa during the era of colonization.[47] Land and necessarily soil are the vital aspects of the earth's physical features, which should be enjoyed by all human beings without restriction. Indeed, some of the problems associated with land tenure systems and soil rights on the African continent are closely linked to the physical and demographic features that exist in a particular country.

It is noteworthy that policies and laws governing land and soil rights in Africa were crafted within the purview of the ideology of patriarchy which potentially edges out the socially and economic weak in regard to land and soil rights. The role of these policies and laws is better understood when the role of courts, which too, are a creature of western colonial rule is made clear. In support of this view, Morris asserts:

> [T]hese courts (in the British colonies) administered basically English law, that is to say the common law, the doctrines of equity and the English statutes of general application in force on a specific date, together with certain Indian Acts, which represented the nineteenth century English law in a codified, and somewhat rationalised, form.[48]

Sadly, the British law as exported to Uganda and other African countries was not all in written form so that it could be understood and applied with certainty. It embodied the unwritten common law and principles of equity, hitherto unknown within the fabric of the African and specifically, the Ugandan society.

The hidden impact of colonial law on land tenurial systems and their regulation of soil rights in the Africa could only be inferred from the language of the written law that incorporated them. Moreover, imported English law sought to edge out African customary law, by including a repugnancy clause in the national law. To date, this ideological legal hangover still informs the laws of many African countries including Uganda. For example the Ugandan Judicature Act provides:

> '[c]ommon law' and 'doctrines of equity' mean those parts of the law of Uganda, other than the written law, the applied law or the customary law, observed and administered by the High Court as the common law and the doctrines of equity respectively.[49]

In furtherance of this ideology, the Judicature Act provides:

> [N]othing in this Act shall deprive the High Court of the right to observe or enforce the observance of, or shall deprive any person of the benefit of, any existing custom, which is not repugnant to natural justice, equity and good conscience and not incompatible either directly or by necessary implication with any written law.[50]

---

[47] Ahmed (1999).

[48] Morris (1970).

[49] The Judicature Act, Cap. 13 (Uganda), section 14 (5).

[50] The Judicature Act, Cap. 13 (Uganda), section 15 (1).

It is noteworthy that within the constructs of the Judicature Act, it would be difficult for an African custom to pass the repugnancy test, and hence its operation is limited to very few circumstances.

What the High Court of Uganda and other courts of Judicature apply is not the African customary law alone; they apply an amalgam of western legal principles as well as the African customary principles, all of which do not favour entrenched rights of women and to some extent non-citizens rights to land and hence soil rights. The problematique in the Ugandan land tenurial systems and their regulation of soil rights, therefore, partly lies in the simultaneous application of the English common law or doctrines of equity and the traditional norms and values of the traditional Ugandan society. Moreover, upon its introduction in the African legal regime, British law sought to subdue and annihilate African customary law, which cannot be distinguished with the way of life of Africans. In this regard, Mugambwa asserts:

> [T]here, recognition of customary land rights was the exception rather than the rule. For example, in Uganda, the British protectorate administration declared most land in the territory Crown land by virtue of the protectorate. Customary land tenure was recognized but within limits. Under the Crown Lands Ordinance 1903, indigenous Ugandans had a right to occupy any land (outside the Buganda kingdom and urban areas) not granted in freehold or leasehold without prior license or consent in accordance with their customary law. However, the Governor had the power to sell or lease such land to any other person without reference to the customary occupants of the land.[51]

Simultaneous application of the hitherto unknown British law in Uganda together with the customary land tenure systems contemporaneously complicate the position of Ugandan women and men, citizens, and non-citizens, in regard to access to soil rights.

Indeed, land, which broadly embodies the soil, is one economically viable resource that all Ugandans, women and men, citizens and non-citizens, have an inherent right to in their individual capacities. That has, however, historically not been the case. For example, in 1845, Marx wrote:

> [W]e do not set out from what men say, imagine, conceive nor from men as narrated, thought of, imagined, conceived, in order to arrive at men in the flesh. We set out from real, active men, and on the basis of their real life process we demonstrate the development of the ideological reflexes and echoes of this life process. Morality, religion, metaphysics, all the rest of ideology and their corresponding forms of consciousness, thus no longer retain the semblance of independence. They have no history, no development: but men, developing their material production and their material intercourse alter, along with their real existence, their thinking and the products of their thinking. Life is not determined by consciousness, but consciousness by life.[52]

It is noteworthy that within the materialistic paradigm purview, men in Africa have historically had better access rights to land. Parpart asserts:

---

[51]Mugambwa (2007).

[52]German (1981).

[I]n the case of Zimbabwe, men continue to have easier access to property than women. In the resettlement schemes set up to provide land to liberation fighters, individual land grants were awarded to men as heads of households. A married woman was, thus, prevented from owning land, and if she were divorced (for whatever reason), she lost the right to stay on the land because it was registered in her husband's name. Only widows and single women could obtain land, and even they had trouble acquiring land because officials were skeptical about their potential productivity.[53]

The western materialistic and patriarchy ideologies have, therefore, worked hand and gloves with some traditional notions of African patriarchy ideologies to discriminate against women and non-citizens in regard to land and soil rights. This aspect is visible in the policies and laws made by contemporary African governments discussed in this chapter.

The friction is now between tradition versus modernity [policies and laws] and legal pluralism, and customary versus statutory laws. Neither of these ideologies works to women's expectations and protection of their rights.[54] Legal pluralism entails various kinds of law such as state law, made by the legislature and enforced by the government. It equally includes religious law, both the written doctrines and accepted religious practice, and customary law, interpretations thereof, and in a globalized world, the project (programme) law. Indeed, in the present globalized era, land law as applied in Africa is conditioned by regulations associated with the western programme of donation.[55] Donations and other forms of foreign aid are intended to serve the interests of developed economies. Donations help to condition African economies to introduce reforms in their traditional land tenurial systems in favour of the donors. The reformed land tenurial systems do not only alienate land [and soil] from the Africans in totality, but equally, alienate the poor, women and men from the soil, which they treasure. In total, none of these legal regimes create an atmosphere of easy access to land and soil in Uganda, much as it is more disadvantageous to the women and non-citizens.

This subsection puts the foundations of land and hence soil legal regime in African into context by highlighting the socio-legal foundations of the land policies and laws as they operate today in Africa. It makes clear that the materialistic nature of the western legal paradigms is essentially at loggerheads with the African customary law, the latter being organic to the African society and hence easily understood and respected. The discussion makes it clear that colonial law and its ideological manifestations is still operational in Africa because it uses the bait of donations to condition the minds of African governments to develop policies and laws that can help maintain the flow of donations.

Donations are not free gifts to African economies; they are only avenues of the neo-colonial western domination and its attendant draining channels of African resources. That trend is disadvantageous to the African economies, it makes them poorer day by day and is incapable of bring about sustainable utilisation of African

---

[53]Parpart (1995).

[54]Sebina-Zziwa (1999).

[55]Ahmed (1999).

land and soils. The critique made using the lenses of African feminism in this and the previous section makes this point clear and will be utilized to further discuss how difficult it is for African economies to achieve sustainable development based on African land and soils under the current land tenurial systems.

4.2   The Link Between Sustainable Development, Land Tenure Systems and Soil Rights

The concept of sustainable development in regard to land and soil rights regime is fairly new. It succinctly appears first in the Brundtland Report in 1987. According to the Brundtland Report:

> Sustainable development is development that meets the needs of the present without compromising the ability of future generations to meet their own needs. It contains within it two key concepts: the concept of 'needs' in particular the essential needs of the world's poor, to which overriding priority should be given; and the idea of limitations imposed by the state of technology and social organization on the environment's ability to meet present and future needs.[56]

The Brundtland Report further provides:

> The essential needs of vast numbers of people in developing countries—food, clothing, shelter, jobs—are not being met, and beyond their basic needs these people have legitimate aspirations for an improved quality of life.[57]

The Brundtland Report provides a basis upon which Uganda's law base to provide for sustainable development.

In the case of Uganda, sustainable development is provided for in the Constitution in 1995 and the post-1995 constitutional reforms of policies and laws. It is noteworthy that the colonial policies and laws regulating land and soil rights did not specifically cater for sustainable development. Key colonial laws that regulated land and soil rights included: the 1884 Land Acquisition Act, the African Order in Council, 1892, the 1899 Land Acquisition Act, 1900 Buganda Agreement, the 1902 Order in Council as amended in 1920, the 1903 Ankole Agreement, the 1903 Toro Agreement and the 1933 Bunyoro Agreement. The impact of the colonial agreements on land tenure systems, and land and soil rights was expressed in the Land Acquisition Act, 1899, which *inter alia* provided:

> In pursuance of the powers conferred by article 3 of the Africa Order in Council, 1892—Section 1: [T]he following enactment of the Governor General of India in Council shall apply to the Protectorate (Uganda), that is to say—The Land Acquisition Act, 1894 (1 of 1894). Section 2: [I]n the application of the said enactment to the Protectorate the following modifications shall be made—(a) where the said enactment provides that any act or thing may or shall be done by the Governor General of India in Council or by a Local Government,

---

[56]World Commission on Environment and Development (1987).
[57]Ibid.

whether with or without the sanction of the Governor General in Council, such act or thing may or shall be done, subject to any directions of the Secretary of State, by the Governor; (b) where the said enactment provides for any notification in any Gazette, such notification shall be made in the official Gazette of Uganda; (e) any land whereof possession is taken under the provisions of the said enactment shall vest absolutely in the Governor for the time being, or, in a trustee or trustees for His majesty, to be appointed by the Secretary of State, who shall have power by order to remove any trustee and appoint any new trustee or trustees.

The Protectorate laws made it clear that the land tenure regimes in the Ugandan Protectorate would be those crafted in Britain and applied first in other colonies such as India. They would then, without much modification, be transplanted into, and superimposed on the African customary laws in the Ugandan Protectorate. The transplant of laws from Britain to Uganda in the described manner created a problematic situation to Ugandans—how would they be applied contemporaneously with the traditional laws in the Protectorate? Attempts by the British colonialists to provide for the modification clause in such laws never addressed any concerns of the Ugandan natives. Those laws remained the oppressive and discriminatory laws crafted within the ambit of the ideology of patriarchy in Britain. Contemporary Ugandan policy and legal principles on sustainable development should, therefore, be viewed with suspicion because not so much has been done to remove the western ideology of patriarchy contestations in the Ugandan land laws as they operate today. At best, such constitutional principles should be viewed as intended to attract the attention of the Western investor communities in the country, while the situation on the ground in regard to access to land and soil rights remains more or less the same as it was more than a century ago. That situation makes it difficult for the African economies in general to receive and integrate modern economic concepts such as sustainable development and implement them with ease in their national land law regimes.

Implementation of the concept of sustainable development, however, is side-lined by the inroads of the western patriarchy ideology in Uganda's policies and its tenets cannot be specifically realized. The desired scenario remains to undo the colonial policies and laws and re-enact them using the lenses of African feminism, which has a motherly affection for the Africans. Repackaging the land policy and legislative norms in African countries is capable of perfecting the land law and hence soil rights regime, and bring about the yearned for sustainable development, which as of now remains only a paper provision. The desire to ensure that the legitimate aspirations of persons in Uganda to land and soil rights and the need to utilize them in a sustainable manner is expressed in the Constitution of Uganda, 1995. Principle XXVII of the National Objectives and Directives of State Policy provides:

(i) [T]he State shall promote sustainable development and public awareness of the need to manage land, air and water resources in a balanced and sustainable manner for the present and future generations. (ii) The utilization of the natural resources of Uganda shall be

managed in such a way as to meet the development and environmental needs of present and
future generations of Ugandans; and, in particular, the State shall take all possible measures
to prevent or minimize damage and destruction to land. . . .

These constitutional principles provide a legitimate basis for critiquing the
Ugandan land tenure regime with a view to establishing whether or not it addresses
the tenets of sustainable development based on land and soil rights. A common
feature of the land tenure, and hence soil rights systems in Uganda is that they are
constituted by both registered and unregistered systems.[58]

Apart from customary land tenure systems, which are the traditional forms of
African land tenure systems, the other forms of land tenure systems were introduced
in the Ugandan legal fabric by the British colonial Government under the colonial
agreements and laws. The latter land tenure systems remain foreign to the African
economies and their impact on the indigenous communities has been to distort the
whole idea of sustainable utilization of land and hence the soils on the continent.
They are commercially oriented, yet commercial interests are difficult to reconcile
with life sustenance concepts. It is probable that if the African system of development of the legal processes had not been hijacked by the forceful entrenchment of the
western ideology through law and colonization, the concept of sustainable development would find a fertile ground upon which to flourish. Unfortunately, that has not
been the case and merely providing for the concept of sustainable development in the
Ugandan and other African policy and legal frameworks in the mentioned circumstances does not guarantees that it will be successfully implemented. Worst of all,
land tenure laws in Uganda purport to provide for equity in matters of access to land,
and hence soil rights, by women and men, citizens and non-citizens, which has never
been the case either in fact or at law. For example, the Constitution of the Republic of
Uganda, 1995 in Article 26 (1) deceptively provides: [E]very person has a right to
own property either individually or in association with others. The phrase 'every
person' is used technically, but deceptively, and within the confines of the exclu-
sionary ideology of patriarchy to portray that all persons in Uganda are equally
protected under the law and are able under the law to access the country's land and
soil rights without any visible constraints. This is of course a fallacy. The reality is
known. Historical societal stratifications constraints grant more rights of access to
land and soil rights to men than to women.[59] Social stratification equally grants or
recognizes the rights of the economically powerful than the poor.

The dichotomy between men and women and rich and poor makes it difficult for
the various strata to work in unison and hence adhere to the principle of sustainable
development.

---

[58]See Section 2 of the Land Act, Cap. 227, which restates the constitutional provisions: [S]ubject to
article 237 of the Constitution, all land in Uganda shall vest in the citizens of Uganda and shall be
owned in accordance with the following land tenure systems—(a) customary; (b) freehold;
(c) mailo; and (d) leasehold.

[59]FIDH (2012).

In recognition of this problem, and in a bid to address it, the Constitution under Article 33(1) provides: [W]omen shall be accorded full and equal dignity of the person with men. By providing for 'according full and equal rights' the Constitution acknowledges first and foremost that there are historical and structural imbalances in women's right of access to soil rights under the various tenurial systems in place in Uganda. Secondly, it insinuates that women must be passive persons in the waiting room to be accorded such rights and dignity, which Ugandan women through their agency may not take with ease. Such a legal construction is intended to distort the motherism paradigm, which if well-articulated, recognizes the right of the mother to bond with the child without externally constructed constraints. Why women be granted the rights and by who?

That is the desire of the ideology of patriarchy, so that they remain at the receiving, purportedly favoured end. In a vague manner, the Constitution under Article 33 (2) provides: '[T]he State shall provide the facilities and opportunities necessary to enhance the welfare of women to enable them to realise their full potential and advancement'.

The Constitution, however, does not put in place specific mechanisms to create opportunities for women to access the land registration services in a much easier form than men or rich, citizens and non-citizens in order to show commitment on the government's undertaking to enhance women's access to land and soil rights. Moreover, the land registration services in Uganda remain expensive and difficult to access largely by the poor.

Indeed, much of the literature on women and land tenure in Africa has viewed the introduction of land titling, registration, and the privatization of land under colonialism and after independence as a setback for women. It leaves women in a state of even greater insecurity with poorer prospects for accessing land [and soil].[60] Walby avers that there are two forms of patriarchy that successively influence lives of women. The first is 'private patriarchy' where women in the home are under the rule of the father, husband or brother. The second is 'public patriarchy' where women enter public spaces of politics but still remain strongly controlled by men.[61] The public patriarchy further closes Ugandan women's chances to have automatic access to the country's land and soil, much as it deceptively appears to open up more avenues for them in that respect.

Other policy frameworks such as the Uganda Vision 2040 equally embody deliberately or inadvertently deceitful provisions purportedly intended to engender sustainable development in the economy based on harnessing the soil. Vision 20140 mission statement provides: '[A] transformed Ugandan society from a peasant to a modern and prosperous country within 30 years'. Transforming Uganda from a peasant to a modern and prosperous economy cannot ignore the fact that Uganda is potentially an agricultural economy that is dependent on how the country's soils are utilized, and by who. Key elements of the Vision 20140 include Uganda

---

[60]Tripp (2004).
[61]Benhabib (1993).

pursuing a planned urbanization policy that will bring about better urban systems that enhance productivity and sustainability while releasing land for commercializing agriculture; and the projection that the country's agricultural productivity will grow at an average rate of about five per cent.

Uganda's Vision 2040 recognizes that agriculture is the main stay of the Ugandan economy employing 65.6 per cent of the labour force and that it contributes twenty one percent to the country's Gross Domestic Product (GDP). It is also recognized under Vision 2040 that agricultural production in Uganda is mainly dominated by smallholder farmers engaged in food and industrial crops, forestry, horticulture, fishing and livestock farming. Agriculture productivity of most crops has been reducing over the last decade mainly due to a number of factors including: high costs of inputs, poor production techniques limited extension services, over dependency on rain fed agriculture, land tenure challenges and limited application of technology and innovation. Vision 2040 equally recognizes that the fertility of Uganda's soil is declining fast and the situation needs to be reversed. Whatever the good intentions of government by coming up with this vital policy document, it remains difficult to achieve sustainable development of the country's economy based on harnessing land. The diverse interests in the ideologically constructed social strata of the rich and poor, women and men, citizens and non-citizens, necessarily cripples any government initiative to promote sustainable development because of lack of a harmonious position and understanding of the development concept by elements of each social stratum.

In such problematic environment, even other government policies, much as they are conspicuously attractive to the readers, cannot be the basis for sustainable development based on harnessing the land. Most of the activities on the land and hence the soil is extractive in nature without the ability and even desire to replenish the depleted soil fertility.

For example, the Uganda Land Use Policy of 2006 whose overall objective is to achieve sustainable and equitable socio-economic development through optimal land management and utilization in Uganda is only attractive on paper but difficult to operationalise. The specific objectives of the Uganda Land Use Policy are: to adopt improved agriculture and other land use systems that will provide lasting benefits for Uganda; reverse and alleviate adverse environmental effects at local, national, regional and global levels; promote land use activities that ensure sustainable utilization and management of environmental, natural and cultural resources for national socio-economic development; ensure planned, environmentally friendly, affordable and well-distributed human settlements for both rural and urban areas; update and harmonize all land use related policies and laws, and strengthen institutional capacity at all levels of Government.

Achievement of the aforementioned policy objectives is pegged on the understanding that land is a fixed resource and is becoming scarce in many areas, and that its ownership has a significant bearing on land use.

The policy, however, uses flowery language that ignores the fact that land in Uganda is owned by the people and government has limited control over it, and hence cannot compel land owners to utilize it in a manner that they do not understand or believe would meet their individual objectives.[62] Moreover, there are observable changing human needs and a growing population resulting in competition of the different uses for the same land and that the demand for land is often greater than its availability.

Consequently, some present land use practices have led to severe land and hence soil degradation even with the Land Use Policy being in place. Other land policies such as the Land Policy, 2013 are equally bedeviled with similar shortfalls and problems. The Vision of the Land Policy is 'a transformed Ugandan society through optimal use and management of land resources, for a prosperous and industrialized economy with a developed services sector'. Its goal is 'to ensure efficient, equitable and optimal utilization and management of Uganda's land resources for poverty reduction, wealth creation, and overall socio-economic development'. It is realised under this Policy, and in conformity with the analysis made herein that this is a postmortem initiative, which can only cure the problem if the land laws are reviewed and revised.

The Policy states among others, that since the advent of colonialism, the country has never had a comprehensive land policy.

What have been in place are the scattered policies and laws on land and soil conservation. Post-independence attempts to settle the land question and deal with fundamental issues in the land tenure systems and land management have been intermittent and limited in scope. For example, the Policy recognizes historical injustices, many of which have resulted in disposition and loss of ancestral access to land and soil rights by some native communities.[63] As a stop gap measure, the government of Uganda came up with a National Gender Policy in 2007 in a bid to mainstream gender concerns in other land related policies.

The priority areas of Gender Policy are improved livelihoods, promotion and protection of rights, participation in decision-making and governance, recognition and promotion of gender in macro-economic management. It is, however, noteworthy that the provisions of the Gender Policy are difficult to read into other policies which were adopted much earlier and without specifically providing for gender concerns in any of them. The problematic situation regarding implementation of land and soil related policies is that not every policy maker in Uganda understands the ability of African feminism to perfect the imperfect situation right from the time of policy formulation through to adoption and enactment of land laws.

Consequently, elements of African feminism are only glossed over in the discussions leading to adoption of key policy documents regarding land tenure systems and soil utilization in the country. They policies and laws that are eventually adopted are, therefore, incapable of taking care of all the land and soil needs of the society. In the result, the link between sustainable development, land tenure systems and soil

---

[62]The Constitution of Uganda under Article 237 (1) provides: (1) Land in Uganda belongs to the citizens of Uganda and shall vest in them in accordance with the land tenure systems provided for in this Constitution.

[63]United Organisation for Batwa (2015).

utilization remains lose and incapable of cementing the relationship between these key parameters.

5   Conclusion

The discussion in this chapter was based on the case of Uganda and used the African feminism perspective, particularly the motherism brand to analyze people's right to land and soil so that these resources can sustainably be utilized. The discussion makes it clear that in Uganda, the concept of land and soil are synonymous, and rights attached to either of them are acquired contemporaneously. Rights to land and hence soil are based on two principles: *jus soli* (soil rights based on birth right citizenship) and *jus sanguinis* (soil rights based on right of blood or familial lineage).

Citizenship, which is one of the criteria upon which land and soil rights are based, is a construct of the legal regime informed by the ideology of patriarchy, which creates social strata of women and men, and citizens and non-citizens and accords each of them different land and soil rights. The land law regime in Africa does not specifically provide for the unconditional *jus soli* (birth right citizenship), which would grant every person born in a particular country automatic citizenship and right to land and soil. The common soil rights in Africa are based on the principle of *jus sanguinis* (right of blood or familial lineage), which must be traced to ascertain a person's right to land and hence soil. The fact that soil access rights in many of African countries are based on the principle of *jus sanguinius*, denies many persons automatic land and soil rights, and hence weakens the link between land and soil rights and sustainable development.

Land policies and laws, including constitutions, constructed within the purview of the western ideology of patriarchy ignore the rights of the African people and cannot be relied upon as the best platform to engender sustainable development in African based on harnessing the land and soils. Conclusively, therefore, without African governments providing an environment where persons in their countries have equal land and soil rights, it remains hypothetical and a fallacy to assume that African soils can be utilized sustainably under the current land tenure legal regime. The only avenue to rectify the situation on the ground is to use the lenses of African feminism perspective, particularly the motherism brand to re-construct policies and re-enact laws related to land ownership and soil utilization so that they laws accord equal land and soil rights to the poor and rich, women and men and citizens and non-citizens in a bid to effectively implement the principle of sustainable land and soil management.

Based on the foregoing conclusions, it is, therefore, recommended that African governments should: review their land regulatory policies and laws, including constitutions, to be able to grant land and soil rights to all Africans based on the principle of *jus soli*. In the case of Uganda, which was used as a case for this chapter, the Judicature Act still requires English common law and doctrines of equity to be applied by Ugandan courts, yet these are difficult to ascertain. It should be amended so that only ascertainable statutory legal principles are applied.

The Constitution should be amended to include an automatic right to Ugandan citizenship for all persons born in Uganda. The *jus sanguinis* principle should only be adhered to where it does not disadvantage any person in regard to land and soil rights in Africa. Future researchers should build on the analysis made herein and step up their advocacy drives to persuade African governments to undertake the necessary reforms in their land regulatory policies and laws.

References

Ahmed N (1999) Race, class and citizenship. The civil rights struggle in Mobile, Alabama Thesis submitted for the Degree of Doctor of Philosophy at the University of Leicester, 1925–85

Asiimwe J (2001) Making women's land rights a reality in Uganda: advocacy for co-ownership by spouses. Yale Hum Rights Dev J 4(1), Article 8. Retrieved March 12, 2018, from http://digitalcommons.law.yale.edu/yhrdlj

Benhabib S (1993) Feminist theory and Hannah Arendt's concept of public space. Hist Hum Sci 6:97–114. Retrieved from https://cpb-us-west-2-juc1ugur1qwqqqo4.stackpathdns.com/campuspress.yale.edu/dist/3/949/fi

Busingye G (2012) Revisiting impediments to Women's land decision-making processes in Uganda, human rights and peace centre, Makerere University. East Afr J Peace Hum Rights 18(2):454

Busingye G (2017) Law and gender relations in land decision-making processes. A case study of Ibanda Town Council, Western Uganda. A Thesis Submitted to The Directorate of Research and Graduate Training of Makerere University for The Award of A Doctor

Chirisa IEW, Mumba A, Dirwai SO (2014) A review of the evolution and trajectory of the African union as an instrument of regional integration. SpringerPlus 3. Retrieved November 14, 2018, from https://www.ncbi.nlm.nih.gov/pmc/articles/PMC3940719

Chopra D, Mülle C (eds) (2016) Connecting perspectives on women's empowerment. Transforming Development Knowledge, IDS Bulletin, 47. Retrieved from https://opendocs.ids.ac.uk/opendocs/bitstream/handle/123456789/9700/IDSB_47_1A_10.1

Coetzee AA (2017) African feminism as decolonizing force. A philosophical exploration of the work of Oyèrónké Oyěwùmí. Ph.D Thesis, Stellenbosch University, South Africa

Ezenwa-Ohaeto N (2015) Fighting patriarchy in Nigerian cultures through children's literature. Stud Lit Lang 10(6):59–66. Retrieved November 20, 2018, from http://www.cscanada.net/index.php/sll/article/viewFile/7217/7576

Food and Agriculture Organization of the United Nations, FAO (2010) Gender dimensions of agricultural and rural employment. Differentiated pathways out of poverty status, trends and gaps, Rome

Gaard G (ed) (1993) Ecofeminism women, animals, nature. Temple University Press, Philadelphia

German L (1981) Theories of patriarchy. International socialism (second series) 12 in 1981. Retrieved March 9, 2018, from http://isj.org.uk/theories-of-patriarchy

Gingrich P (1999) Marx's theory of social class and class structure. University of Regina, Department of Sociology and Social Studies. Retrieved November 14, 2018, from http://www.u.arizona.edu/~gradisek/STORIESmarx.html

Government of the Republic of Uganda Ministry of Lands, Housing and Urban Development, (2006) The Uganda National Land Use Policy, 2006. Retrieved from mlhud: http://mlhud.go.ug/wp-content/uploads/2013/08/National-Land-use-Policy.pdf

Government of the Republic of Uganda, Ministry of Lands, Housing and Urban Development (2013) The Uganda Land Policy, 2013. Retrieved from http://extwprlegs1.fao.org/docs/pdf/uga163420.pdf

Hartmann H (1979) The Unhappy Marriage of Marxism and Feminism, Capital and Class 1979 Summer

International Federation for Human Rights, FIDH (2012) Women's rights in Uganda. Gaps between policy and practice. Retrieved March 15, 2018, from https://www.fidh.org/IMG/pdf/uganda582afinal.pdf

Kagoda AM (2008) The effect of land tenure system on women's knowledge-base and resource management in Manjiya County, Uganda. Educ Res Rev 3(12):358–335. Retrieved from http://www.academicjournals.org/ERR

Kandiyoti D (1988) Bargaining with patriarchy, gender and society. Special Issue to Honour Jessie Bernard, vol 2(3), pp 274–290. Retrieved from http://links.jstor.org/sici?sici=0891-2432%28198809%292%3A3%3C274%3ABWP%3E2.0.CO%3B2

Mays TM (2003) The Gregg Centre for the Study of War and Society. J Confl Stud 23(1). Retrieved March 9th, 2018 from: https://journals.lib.unb.ca/index.php/jcs/article/view/353/552

Morris HF (1970) Some perspectives of the East Africa legal history. Crime in East Africa: 3. The Scandinavian Institute of African Studies

Mugambwa JT (2007) A comparative analysis of land tenure law reform in Uganda and Papua New Guinea. J South Pac Law 11(1). Retrieved March 15th, 2015, from http://www.paclii.org/journals/fJSPL/vol11no1/pdf/Mugambwa.pdf

Muinde DK (2013) Assessing the effects of land tenure on urban developments in Kampala. Enschede, The Netherlands March 2018

Mwale PN (2002) Where is the foundation of African gender? The case of Malawi. Nordic J Afr Stud 11(1):114–137. Retrieved March 14, 2018, from http://www.njas.helsinki.fi/pdf-files/vol11num1/mwale.pdf

Namubiru-Mwaura E (2014) Land tenure and gender. Approaches and challenges for strengthening rural women's land rights. Women's Voice, Agency, & Participation Research Series 2014 No. 6. The World Bank

Napikoski L, Lewis J (2018) Patriarchal society according to feminism. Feminist theories of patriarchy. Retrieved November 14, 2018, from https://www.thoughtco.com/patriarchal-society-feminism-definition-3528978

Nkealah N (2016) West African feminisms and their challenges. J Lit Stud 32 (2). Retrieved March 7, 2018, from https://www.tandfonline.com/doi/abs/10.1080/02564718.2016.1198156?journalCode=rjls20

Okonya JS, Kroschel J (2014) Gender differences in access and use of selected productive resources among sweet potato farmers in Uganda. Agriculture and Food Security. Retrieved from https://agricultureandfoodsecurity.biomedcentral.com/articles/10.11

Parpart JL (1995) Gender, patriarchy and development in Africa. The Zimbabwean Case, Working Paper #254, Dalhousie University. Retrieved March 2018, retrieved from http://gencen.isp.msu.edu/files/6914/5202/7078/WP254.pdf

Saunders P (1990) Social class and stratification. Routledge, New York

Sebina-Zziwa A (1999) The paradox of tradition. Gender, land and inheritance rights among the Baganda, Unpublished Doctor of Philosophy (Ph.D) dissertation, University of Copenhagen

Slavchevska V, De la O Campos AP, Brunelli C (2016) Beyond ownership: women's and men's land rights in sub-Saharan Africa, Food and Agriculture Organization of the United Nations, Working Paper. Annual Bank Conference. Retrieved from: http://pubdocs.worldbank.org/en/170131495654694482/A2-ABCA-Slavcheska-et-al-2016-Beyond-ownership-working-paper.pdf

The World Commission on Environment and Development (1987) Report of the World Commission on Environment and Development: our common future. Retrieved from: https://sustainabledevelopment.un.org/content/documents/5987our-common-future.pdf

Tripp AM (2004) Women's movements, customary law, and land rights in Africa. The case of Uganda, African Studies Quarterly. Retrieved from http://www.africa.ufl.edu/asq/v7/v7i4a1.html

United Nations, General Assembly (1995) Beijing declaration and platform for action. Beijing + 5 Political Declaration and Outcome. Available at: https://www.unwomen.org/-/media/head quarters/attachments/sections/csw/pfa_e_final_web.pdf?la=en&vs=800

United Organization for Batwa Development in Uganda (UOBDU) et al (2015) Indigenous peoples in Uganda: a review of the human rights situation of the Batwa people, the Benet people and pastoralist communities. Alternative report to the Initial report of the Republic of Uganda, 55th session of the United Nations Committee on Economic, Social and Cultural Rights 1st–19th June 2015. Retrieved from http://www.forestpeoples.org/sites/fpp/files/publication/2015/04/080515-alternative-ngo-report-cescr-uganda.pdf

Valledor-Lukey VV (2012) Pagkababae at Pagkalalake (Femininity and Masculinity). Developing a Filipino Gender Trait Inventory and predicting self-esteem and sexism. Child and Family Studies – Dissertations, Syracuse University

Walker C (1995) Conceptualizing motherhood in twentieth century South Africa. J South Afr Stud 21:417–437. Retrieved from https://www.jstor.org/stable/2637252?seq=1#page_scan_tab_contents

Wily LA (2011) The status of customary land rights in Africa today rights to resources in crisis. Reviewing the Fate of Customary Tenure in Africa - Brief #4 of 5. Retrieved from https://rmportal.net/library/content/status-of-customary-land-rights-rights-to-resources/at_download/file

Soil Governance and Sustainable Land Use System in Nigeria: The Paradox of Inequalities, Natural Resource Conflict and Ecological Diversity in a Federal System

Bibobra Bello Orubebe

1 Introduction

The Federal Republic of Nigeria comprises an area of approximately 923,853 square kilometres.[1] It is bounded to the north by the African nations Niger and Chad, to the west by Benin Republic, to the east by Cameron; to the south by Sao Tome and Principe, and Equatorial Guinea. Nigeria's population is currently estimated at over 180 million people.[2] There are approximately three hundred ethnic groups of which the Hausa-Fulani, Yoruba and Igbo are 'politics engineered' majority ethnic groups.[3] Nigeria's climate and ecology is diverse, with the Sahel desert close in the north, tropical forest in the south, mountains in the east and mangrove swamps in the core Niger Delta where the River Niger meets the Atlantic Ocean.

The North is predominately Muslim with indigenous minority Christians, while the southern Nigerians are essentially Christians with a sizeable Muslim Population. Geographically, Nigeria is situated between 4° and 14° north of the equator and longitudes 3° and 15° east of the International Greenwich Meridian. It enjoys a diversity of climates and marked environmental features, including oil and gas-rich and sensitive wetlands. The country consists of several extensive physiographical plateau surfaces including the Jos Plateau, the Udi Plateau, the Manbila Plateau and the North-Central High Plains. Today, the country is divided into 36 states,[4]

---

[1]Udo (1970). Note that in the 1970s the official measurement used in Nigeria was 356,700 square miles. This figure was converted to kilometers by the author for easy reference and comprehension.

[2]Nigeria National Population Commission, NPC (1998). The current figure is an estimate out of population figure projections.

[3]NPC (1991). The current figure is an estimate out of population figure projections.

[4]Constitution of the Republic of Nigeria (1999), Section 3.

B. B. Orubebe (✉)
College of Law, Novena University, Ogume, Nigeria

© Springer Nature Switzerland AG 2020
H. Yahyah et al. (eds.), *Legal Instruments for Sustainable Soil Management in Africa*, International Yearbook of Soil Law and Policy,
https://doi.org/10.1007/978-3-030-36004-7_9

774 local government council areas and a federal capital territory with three branches[5]—the Executive (Section 5), the Legislature (Section 4) and the Judiciary (Section 6) of the Constitution of Nigeria (1999 as amended), and three tiers of Government.[6] Though the present governance structure may appear to be modern or democratic, a close analysis reveals its diverse historical, military and ethnic influences, including the impacts of colonialism, post-colonialism and post-independence ethnic politics.

The effect of the inequalities associated with the Nigerian federal system since the colonial era has tremendously impacted on the soil administration in the country. Consequently, there is a persistent distrust among various ethnic groups and the government with regards to land in particular, and the allocation of natural resources in general. Although the complications of soil governance could be traced to the colonizers, they have become more glaring with the promulgation of the Land Use Act of 1978. This is a paradigm shift from the traditional system of land ownership which recognised individuals, family or the community as the land owners. The law has given the state government the right to administer the land in trust within its jurisdiction. Previously, the land was controlled by the family heads, traditional rulers, communities, etc. There is now conflict of interest among states, ethnic nationalities, communities, families, and individuals. The law appears to have done more harm than good.

The situation is exacerbated by the negative impact of climate change and global warming, a fact that has necessitated an unprecedented migration of people and livestock to other parts of the country. This movement has resonated the people's concern towards their heritage (land). This is crucial because so much of life depends on it. Despite resistance from the traditional land owners, the suspicious intruders too are resolute to occupy the land by any means possible, sometimes exploring strenuous ethnic, religious, and other influences of government authority and reliance on federal law. It has now become a matter of survival of the fittest. This is a key natural resource access conflict that is confronting Nigeria. In effect, the government, too, seems to be helpless, because it cannot protect the land without the individuals or the communities. Paradoxically, it is the indigenes that actually protect the land for the state, whereas the law gives the ownership right to the state.

It is the thrust of this chapter to unravel these contradictions that have hindered the success of soil governance in Nigeria. This chapter additionally attempts to suggest ways of extricating the country from this monster.

This chapter is subdivided into eight sections. Section 1 is an overview of the issues discussed in the chapter. Section 2 attempts to define, describe and explain the basic terms used in the chapter. This is important considering the elusive nature of words. Notwithstanding, concepts used in the chapter are given both their common and contextual explanations. Section 3 exposes the complexities in the Nigerian

---

[5]The Executive (Section 5), Legislature (Section 4) and the Judiciary (Section 6) of the Constitution.

[6]The Executive (section 5) Legislature (Section 4) and the Judiciary (Section 6) of the Constitution.

socio-political and economic configuration vis-à-vis soil governance. It analyses the remote and immediate causes of the soil induced conflicts that have ravaged the country in recent times. Section 4 deals with cases bordering on traditional land-ownership as opposed to the Land Use Act of 1978. The two regimes are juxtaposed and evaluated for a clearer understanding, as well as linking the present soil governance system and its attendant challenge to the Act. Section 5 puts together the legal provisions in the Nigerian law that considers soil protection. This section also identifies the inadequacies inherent in the available provisions that have contributed to the country's soil governance woes. Section 6 is a consideration of the international legal framework on soil governance. It examines the global concern and advocacy for the sustainable and productive use of soil. It further makes observations on some regional and global instruments pertaining to sustainable soil management. Section 7 contains recommendations that are geared towards achieving zero net soil degradation in Nigeria. The suggestions provided here are so holistic and apt that, with proper and meticulous consideration and political will, they can turn the morbid state of unsustainable soil governance in the country around. Finally, Sect. 8 is a recap of the chapter.

2 Definition of Terms and Key Concepts

This section deals with some key words with terminological implications which need to be defined or explained. Often times, the meanings of words are being misconceived due to shared similarities or dual perspectives. These words include *land, soil, farmer, herdsman, soil governance, soil management, and land degradation neutrality*. It is pertinent to note that the Nigerian Land Use Act of 1978, which is regarded as the central statute which laid out the legal and governance structure of land, does not define in lucid terms what constitutes land or soil. Nevertheless, in this chapter, *land* is used in both its juristic and real property meaning as a three-dimensional part of the Earth's surface which is an immovable and indestructible[7] space not covered by water. It includes the ground or soil[8] and extends to the space above and below. *Soil*, on the other hand, refers to a mixture of organic matter minerals, gases, liquids, and organisms that together support life.[9] In view of the above, land and soil can be used interchangeably.

Interestingly, two critical uses of land in Nigeria are farming and animal husbandry—cattle rearing. Although farming may involve crop as well as animal production, in this chapter it refers to the former as this is normally the case in Nigeria. In other words, farming is the act of cultivating the land for crop production. Accordingly, a *farmer* is one who engages in this act. He promotes and improves the

---

[7]Garner (2015).

[8]Allaby and Park (2017).

[9]Ponge (2015).

growth of crops. On the other hand, a *herdsman* is one who lives a nomadic life, caring for animals (cattle) in places where these animals wander pasture lands. In the Nigerian context, herdsmen are mostly associated with the Fulani ethnic group. Therefore, they are referred to as Fulani Herdsmen.

Another critical term to be considered is *land degradation neutrality*. This according to UNCCD COP 12 is "*a state whereby the amount and quality of land resources, necessary to support ecosystem functions and services, and enhance food security, remain stable or increases within specified temporal and spatial scales and ecosystems*".[10] However, it is conceived in this chapter as agricultural and other uses of land including restoration of degraded natural and semi-natural ecosystems that provide vital, direct and indirect means of livelihood to people and working landscapes without compromising its present and future quality. The general principles, aims, objectives and structure of acceptable soil governance are contained in the United Nations Convention to Combat Desertification (UNCCD), which Nigeria and many African countries are signatories.

Notably, governance of the soil tends to attract more concern in recent times as nations and international bodies discuss it at various meetings and conferences. Nevertheless, it has not yielded the desired result where policy-makers would streamline it separately like the climate change scenario. In fact, legal or policy issues on soil are currently dealt with or embedded in the UNCCD and other related documents (instruments). However, this is an acknowledgment of the significant role played by soil in the affairs of humanity. Conspicuously, it is quite impossible to tackle environmental issues such as biodiversity, climate change, etc. without a response to soil sustainability. According to Ben Boer and Ian Hannam, "While biodiversity loss and climate change have garnered close attention, issues of land degradation and sustainability of soil has attracted less focus in international forums and by national governments."[11]

Nevertheless, soil governance seems to be more focused on agricultural perspective due to the increased awareness attached to the potential risks and the concern that food insecurity poses to most regions of the world. Nigeria is a member of the Global Soil Partnership (GSP) by virtue of its membership of the Food and Agriculture Organization (FAO). FAO and its members initiated GSP with a view to improving governance of the limited soil resources of the planet, in order to guarantee healthy and productive soils for a food-secured world, as well as supporting other essential ecosystem services.

This said, it is safe to opine that *soil governance,* in the context of this chapter, refers to the laws, policies, strategies, and other processes of decision-making employed by the Nigerian State or Government regarding the use of soil,[12] developed with the active participation of the citizenry that includes restoration or remediation of degraded soils, capable of holding violators accountable, and afford

---

[10]UNCCD COP 12 (2018), p. 10.

[11]Boer and Hannam (2015).

[12]Junge et al. (2010).

citizens unfettered access of redress through the courts.[13] Unfortunately, in Nigeria, soil governance is biased against promoting sustainable agriculture and ensuring food security. In terms of legal implications under the Nigerian context, governance of the soil or land differs from "land" or "soil" management. Thus, soil governance entails national, regional and international collaboration between governments, the private sector and individuals. The aim is to attain thorough implementation and enforcement of coherent policies that encourage practices and methodologies that regulate usage of the soil or land as a natural resource and to avoid conflict between users, as well as to promote sustainable land management.[14] Similarly, *soil management*, involves usages including techniques that ultimately increase and maintain soil integrity in terms of fertility, output in relation to agricultural yield and carbon sequestration as conceived under the Paris Agreement on climate change.

3 Nigerian Realities

3.1 Political Dimensions to Soil Governance Conflict

Before the advent of British colonial rule, Nigeria was comprised of chiefdoms, kingdoms and autonomous ethnic societies with separate political and customary institutions. The British initially created three separate political and economic entities—Northern, Southern and the entity administered by the Royal Niger Company (RNC) called the Oil Protectorate. The north had its administrative headquarters in Lokoja, while the headquarters of the south was in Forcados. Faced with the reality of consistent budget shortfalls in the North and surpluses in the South, the British amalgamated the three areas for administrative convenience in 1914, with Lord Lugard, the first colonial Governor-General, extending his dual mandate to the entire area which he called Nigeria.

At the end of the Second World War and with the establishment of the United Nations, the British colonial government was pressured to grant self-rule and independence to Nigeria. In 1960, Britain installed a northern dominated government which in turn manipulated the result of the 1962 census for ethnic reasons. According to Professor Cathering S.M. Duggan[15]:

> ... When the census results were released this manipulation was obvious, and another census had to be taken in 1962 which was also highly politicized. The census was not the only count subject to manipulation; the 1964 elections were also filled with voter rigging and intimidation. The government was overthrown on January 15, 1966, when a group of predominately Eastern army officers killed the Prime Minister and several ministers.

---

[13]Orubebe (2017), p. 51.
[14]UNEP (2016), p. 33.
[15]Duggan (2009).

After days of violence and uncertainty, the president of the senate invited the head of the army, Maj. Gen. Johnson Aguiyi-Ironsi to resume power as head of the government. Many northerners feared that Aguiyi-Ironsi, an easterner was simply trying to shift the regional power balance previously in favour of the North. Northern Nigeria erupted in deadly riots, with widespread attacks on Southerners, particularly, easterners (Ibos) living in the northern region. . .[16]

Consequently, in July, northern soldiers murdered Aguiyi Ironsi and enthroned a young northerner, Lt. Col. Yakubu Gowon. The attacks in the north continued, and as many as 100,000 eastern Ibos were killed, about 500,000 displaced in a matter of few months.[17] Therefore, the eastern regional government declared the situation untenable and seceded from Nigeria in May 1967, renaming itself the Republic of Biafra. Invariably, a civil war broke out and as many as three million easterners were reportedly killed.

On July 30, 1975, a group of Hausa Fulani junior northern army officers seized power in a peaceful coup d'état while Gowon was out of the country. The coup leaders installed their fellow Hausa-Fulani northern officer, Gen. Murtala Muhammed who continued the Hausa-Fulani northern agenda and moved the political capital of Nigeria from Lagos to Abuja in Northern Nigeria. Following Gowon's example, he split the country's twelve states to nineteen, ostensibly to strengthen the northern Hausa-Fulani power base in the Nigerian Federation. Six months after taking office, General Muhammed was assassinated in a coup attempt. Lt. Gen. Olusegun Obasanjo, a western war hero, took over as the Head of the Supreme Military Council and Head of State of the Federal Republic of Nigeria. He continued with Muhammed's agenda, and empaneled selected Nigerians to draft a constitution after which he supplanted the Land Use Decree of 1978 which was not in the agenda of the Constituent Assembly, but was surreptitiously added to the constitution of Nigeria in 1979 without due process. As will be demonstrated subsequently in this chapter, the above geopolitical and historical perspective of Nigeria has impacted exceedingly on the soil governance system, which is in effect the bane of major inter-ethnic conflicts in the country.

3.2   Ethnic and Economic Dimensions to Soil Governance Conflict

Nigeria like most African countries is facing the problem of food security, degradation of land, desertification, pollution, and creation of wasteland. In practical terms, these problems have led to mass migration of the Fulani herders, who are Muslims, down to the southern part of the country due to insufficiency of grass in their home land to sustain their livestock. In the cause of their movement, their cattle often

---

[16]Duggan (2009).
[17]Osaghae (1998).

destroy farms owned by the indigenous people who are mainly Christians. With deep-seated prejudice, the Fulani herdsmen find it difficult to appreciate the customary land governance system of the other ethnic groups. This, coupled with their pastoral nomadic habit, brings them in conflict with the indigenous land owners in the Middle Belt and the southern states of the country. In some instances, the cattle eat up crops and when the owners of the crops react, the herdsmen instead of being apologetic, often lay claim to the ownership of the land on which the crops are planted.

The herders contend that grass belongs to Allah[18] and as such the crops are forms of grass owned by everybody since Allah is the rightful owner. Ironically, they never view their cattle as Allah's property that could be collectively owned by everybody. Besides, they understand land to be owned by the government under the Land Use Act.[19] On the other hand, the Christian farmers of southern Nigeria insist that their customary soil or land governance recognises customary ownership including Kola or tribute land tenancy system.[20] Obviously, the customary ownership of land tends to be more successful. In fact, before the advent of colonialism, there seemed to be less ethnic clashes on account of land.

Notwithstanding, the above contention has snowballed to major crises in several cases. In reality, the Fulani herders travel from the far northern states to the southern states in search of good pasture for grazing. Instead of negotiating with the owners, they claim to be co-owners of the land. Apparently, the reaction of the traditional owners has not been friendly either, especially, when cash or food crops on farm lands are destroyed without any form of compensation. In addition to the aforementioned, these herdsmen are so determined and desperate for the survival of their animals that some of them maim, kill or massacre the farmers that attempt to prevent the cows from grazing on the land and farms without the customary kola, tribute or compensation. A worst-case scenario is when some Fulani herders would burn down farms in order to yield green pasture for their cattle months later.[21] This situation explains why states like Ekiti banned open grazing. The Ekiti State law, although not perfect, criminalises acts that degrade fertility of soil or land belonging to another person by a herdsman, either by burning or destroying crops in farms.[22]

---

[18]Agbosu (1988).

[19]Section 1 of the Land Use Act, Cap. L5 LFN (2004).

[20]Kola tenancy is defined under the Kola Tenancies Law, Cap. 69, Laws of Eastern Nigeria (1963), as "a right to the use and occupation of land which is enjoyed by any native in virtue of a kola or other token payment made by such a native or any predecessor in title in virtue of a grant for which no payment in money or in kind was exacted."

[21]PM News Nigeria (February 12th, 2018).

[22]Prohibition of Cattle and other Ruminant grazing in Ekiti (2016). This law criminalises grazing in some places within the state and outside certain periods in the day. It also prohibited the carriage of any kind of weapon by herdsmen (Prohibition of Cattle and other Ruminant Grazing in Ekiti, 2018).

The Fulani herdsmen and farmers crisis are further compounded by the arguments for and against cattle colonies or ranches. While many view ranches as the panacea to the crisis, others particularlly the federal government suggest the establishment of cattle colonies as the solution, according to the protagonists of this view point the establishment of the cattle colonies is a measure to 'quickly curb the incessant bloody clashes between farmers and herdsmen'.[23] At some point, the two terms are mixed up and become confusing to even the legislature[24] and the executive. Whereas a ranch is an area of land with equipment and structures designed for grazing and the raising of livestock such as cattle for meat or wool, a cattle colony is essentially set up for the purpose of selling of cattle and beef products. A colony is a market square for cattle meat. The debate became more intense when the federal government indicated interest to establish 'colonies' instead of ranches in sixteen states of the federation.

Some critics queried the federal government's interest in the matter, as it is the states that own the lands in their territories by the provisions of the Act. According to Nigeria's foremost constitutional law expert, Professor Ben O. Nwabueze:

> The cattle colonies which the federal government proposes to establish in every state of the federation can, therefore, mean nothing other than a place for the settlement of Fulani Herdsmen...Its character as a place for the settlement of Fulani herdsmen is implicit in the agriculture minister's long presentation giving details of the proposed project....[25]

Nwabueze's use of the term 'colonies' instead of ranches were informed by the Minister of Agriculture's misuse of the word in his proposal. However, from the above quote, he has exposed the interest and politics behind the term. Rearing of cattle is supposed to be a private business, and, as such, it is not the government's responsibility to establish either ranches or colonies for the herders. This is more so because the president is himself a Fulani and many Nigerians impute ethnic connotations to this rather invidious proposal.

In all of these, the Fulani herdsmen are neither interested in ranches nor colony. Their interest is the traditional nomadic pattern which is their way of life from antiquity. One of their strong points of argument is that cattle routes which were provided for them by the colonial government have been trespassed by farmers and other land developers in the Middle Belt and other parts of the country. Therefore, they are prepared to preserve their perceived heritage in whatever circumstance. To make good their vow, some of them have started carrying AK-47 rifles[26] to ward off any opposing farmer. A disastrous dimension to this whole armed conflict is that the law enforcement agencies of the Federal Republic of Nigeria do not take proactive steps to prevent the killings. A situation that led to Chief E. K. Clark, a First Republic Minister of Information, senator, and elder statesman, challenging the government of

---

[23]See Statement of Chief Audu Ogbeh, the Minister of Agriculture and Rural Development.

[24]Daily Post (January 16th, 2018a).

[25]PM News Nigeria (January 30th, 2018).

[26]Infra note 44.

the day to "make arrest of herdsmen carrying AK-47 rifles illegally across Nigeria". Barely four days after the elder statesman's charge, the Nigerian police arrested a herdsman with AK-47 rifle in Enugu state.[27] This is surprisingly terrifying because the traditional herdsmen were never known to be carrying guns. These men have sent countless men, women and children to their early graves because of land dispute. They sack villages, thereby rendering helpless village widows, widowers and orphans. Although the impact of herdsmen killings is like an 'albatross on the neck' of the whole nation, Benue state appears to be the epicentre.[28] In the past, Miyetti Allah and other "Fulani leaders openly claimed responsibility for the killings, but alleged being provoked by wanton rustling of their livestock."[29]

In view of the above, critics blame the government in power for being bereft of solution to the crisis. The issue is even further aggravated by the steps taken by Benue and Taraba state governments to help their citizens from the "claws of these marauding beasts".

Ironically, the situation has become even more horrible. In this regard, the Inspector General of Police blamed the spate of killings on the signing into law of the Anti-Open Grazing Bill by the two states. If the assertion of the Inspector General of Police is anything to go by, then what about the similar situations in Plateau, Nassarawa and Kogi states where the anti-grazing law do not exist but the wanton killings by Fulani herdsmen have continued unabated? As it is now, Nigerians believe there is "more to these killings than what meet the eyes". Obviously, there is a political and ethno-religious dimension or undertone to it. Perhaps that was what General TY Danjuma (Rtd), a former minister of defense, saw when he vociferously urged Nigerians on a national television to "resort to self-help or you all die."[30] He made it clear that the security agents colluded with the Fulani herdsmen to kill people from other tribes.[31] So, does this mean ethnic cleansing? If this is the purpose for all the killings, why is the cleansing? Is the intention to flush out the Tivs and other minorities in order to claim their lands? These and many more are questions begging for answers in the Nigerian current soil governance crises. Danjuma's position was supported by Prof. Sagay who took it from the point of view the law that every citizen has "the right to self-defence".[32] Danjuma was not alone; elder statesmen like Chief Olusegun Obasanjo, Wole Soyinka, Pastor Tunde Bakare and a host of others including the US President, Donald Trump observing that:

...we have had very serious problems with Christians who have been murdered, killed in Nigeria. We are going to be working on the problem....because we can't allow that to happen.[33]

---

[27] Pulse NG (July 4th, 2018).

[28] Punch Newspapers (May 2nd, 2018b).

[29] The Guardian Nigeria (April 13th, 2018).

[30] Vanguard (March 24th, 2018b).

[31] Vanguard (March 24th, 2018b).

[32] Daily Post (March 26th, 2018b).

[33] Quartz Africa (May 1st, 2018).

The spate of killings in the Middle Belt and elsewhere is unimaginable. On 24 April 2018, the Fulani herdsmen killed two priests (Rev Fathers Gor Joseph and Felix Tyoloha) and 17 parishioners in a Catholic Church in Benue state.[34] The killers also burnt down the church. It is against this pathetic background that Nobel Laureate, Prof. Wole Soyinka reiterated his view that:

> Nigeria is sliding from ethnic cleansing to genocide. The plain expression is ethnic cleansing and we must not beat around the bush. The shade of Rwanda hangs over the nation (Nigeria). . . .[35]

The Benue youths, seemingly tired of the trend, allegedly carried out a reprisal attack on Huasa-Fulani settlements within their territory.[36] This confirmed the fears of many Nigerians and international observers that the Nigerian crisis was turning out to be an ethno-religious conflict thereby downplaying the real cause of the crisis which is soil governance or access to soil resources.

Despite the outcry on these killings, according to Senator Eyninaya Abaribe,[37] the Commander-in-Chief who is also a Fulani, claims not to be aware of all the Fulani herdsmen killings happening in the country.

He further moved that "President Buhari is incompetent." Similarly, many critics have urged the President to proscribe the Fulani herdsmen as terrorists, but he pays deaf ear to them. However, the Middle Belt Forum (MBF) insists that such a declaration will purge the President of bias. In a message to the people of Benue state, President Buhari, contrary to the expectation of the bereaved Benue people, enjoined them to "accommodate the killer herdsmen as their brothers".[38] Meanwhile, President Buhari in response to comments by the Archbishop of Canterbury, His Grace, Justin Welby, while in London, the United Kingdom, said:

> The problem is even older than us. . . It has always been there, but now made worse by the influx of gunmen from the Sahel region into different parts of the West African sub-region. . .these gunmen were trained and armed by Muammar Gaddafi of Libya. When he was killed, the gunmen escaped with their arms. We encountered some of them fighting with Boko Haram. . . .Herdsmen that we used to know carried only sticks and maybe a cutlass to clear the way, but these ones now carry sophisticated weapons. The problem is not religious, but sociological and economic. But we are working on solutions.[39]

In effect, the herdsmen and farmers clashes have hyped the already critical food insecurity situation in the country. This is significant in the sense that the states most affected are major players in crop production. For instance, Benue is nicknamed the "Food Basket of the Nation" because of this role. Quite unfortunately, 'things are falling apart.' Farmers can no longer work freely and effectively in their farms for the fear of unsuspected herdsmen attacks. Another possible problem originating from

---

[34]Daily Post (April 24th, 2018c).

[35]Premium Times (April 30th, 2018b).

[36]The Cable Nigeria (April 27th, 2018).

[37]Vanguard (April 13th, 2018c).

[38]Punch Newspaper (January 15th, 2018a).

[39]Premium Times Nigeria (April 12th, 2018a).

this crisis is the call for disintegration. Many indigenous people of the Middle Belt view secession as the solution to the incessant killings in the region.

After decades of social, economic and political crises, and decline, it is more apparent than real to posit that the next frontier of crisis in Nigeria is soil governance or land tenure crisis. This is more frightening because the root cause of this new challenge, which is balanced access to land resources and lack of adaptation to the deleterious effects of climate change, is greatly misconceived by the leaders of the country. The problem of access to soil or land resources is central to the nature of conflicts and co-operation between ethnic groups. It is easy to point out that desertification in the far northern Nigeria, particularly in the homelands of the Fulani herdsmen, which occasions their mass migration with their animals to other regions illustrates the changing modes of access to soil and pasture. Consequently, this has culminated to the present-day tensions and conflict between the Muslim Fulani herdsmen and the Christian indigenous farming groups. The Nigerian government's efforts to provide mutually benefiting solutions capable of addressing stakeholders' interests so far have exasperated social conflicts and violence, rather than creating security for all groups concerned. Certainly, there is mistrust among the various ethnic groups, and at the moment, there seems to be no lasting solution to the crises.

This is as a result of the inadequacy of the current legal and governance framework, institutional barriers to sustainable soil governance, and the need for international soil governance.

4   Legal Observations on Soil Governance in Nigeria

Before the promulgation of the Land Use Act of 1978 by the then military junta, soil governance was essentially based on customary rules in which the head of the community or family held and administered the land in trust for his people. In this regard, his consent must be sought before deposition of land[40] could be made. Nigeria promulgated the Land Use Decree[41] in 1978, a time the northern states were subject to the Land Tenure Law of 1962.[42] However, in southern Nigeria, the rules of customary law that applied varied from place to place. Thus, the law relating to lands and indeed soil(s) underwent changes so dramatically that it impacted on the cultures of about 350 diverse ethnic nationalities.[43] A direct consequence of these changes is that customary property rights on soil or land have diminished in several parts of the country, particularly in the southern states. Accordingly, individual property rights of access to use of soils, land and other natural resources thereon[44]

---

[40] Agbosu (1988).

[41] Nigerian Land Use Act LFN Cap L5 (2004).

[42] Adigun and Omotola (1982).

[43] Adigun and Omotola (1982).

[44] Situma (2003).

prevalent under customary law were changed to a dubious government owned trust styled land rights regime in direct contrast to the dictates of modern Nigerian individualistic and socio-economic reality.

Under customary law, soil governance refers to traditional principles, policies, beliefs, strategies, and the processes of decision-making that guaranteed its sustainable and agreed uses, fertility and general productivity that met the people's food security and cultural needs. Thus, soil governance ensured that the many uses of soil were met without conflict. In other words, traditional soil governance mechanisms were usually encapsulated around the belief systems of the people and remained the most essential resource to the people. In fact, it was perceived as their heritage. This was epitomised by a complex, yet balanced, individual, family or communal interest with the head at the apex of the governance structure. This traditional or customary context of overall land governance addressed the peoples' livelihood, cultural needs of soil and other alien temporary users' interest which included life habits such as pasturing domesticated herds of cows, goats, sheep, donkeys, etc.

Another intriguing feature of the customary soil/land governance system in Nigeria (particularly in southern Nigeria) was that it promoted complex and distinct livelihoods such as agriculture, diverse cultural and economic activities and, above all, ensured balanced access to soil resources. These features of the customary soil or land governance system, as it relates to the basis of ownership of land under customary law, were not easily understood by the colonial authorities.

In fact, in the hallmark case of *Amodu V. Secretary of Southern Nigeria,* Lord Haldane made the following observations about the customary land governance system or tenure:

> ...the next fact to bear in mind in order to understand native land law is that the notion of individual ownership is quite foreign to native ideas. Land or soil belongs to the community. All the members of the community, village or family have an equal right to the land but in every case the chief or headman of the village or community or head of the family has charge of the land and in loose mode of speech is sometimes called the 'owner'. He is to some extent in the position of a trustee and as such holds the land for the use of the community or family. He has control of it and any member who wants a portion of it to cultivate or build upon goes to him for it. But the land or soil or resource so given remains the property of the community or family.[45]

Reacting to this lack of understanding, as noted above, Omotola,[46] Coker,[47] Elias[48] and other experts[49] view it as a mischaracterisation of land or soil governance by Lord Haldane as it relates to family or communal ownership. Although the positions of the three jurists are well thought out from two competing perspectives, suffice to say that the joint ownership of soils/land governance system under

---

[45]Amodu v. Secretary of Southern Nigeria (1921) A.C. 399–404.

[46]Adigun and Omotola (1982).

[47]Coker (1966).

[48]Elias (1971).

[49]Among the leading experts on African Land Law are: Allott, Woodman, Gordon, Bentsi-Enchill, Nwabuaze, Kludze, Asante, Ollennu, Obi, Olayede, Ezejiofor and others.

customary law represents the culture of the people of southern Nigeria who were subject to it.[50] The principle of joint ownership was so pertinent that the communal or family head's consent was essential to any disposition of soil/land.[51] The role or position of the community or family "head" was so revered under customary land governance system that he could even resist moves by other family members who attempted to sue him to give an account.[52] Another interesting principle under customary land governance was the well-established fact that no member of the family had any alienable right in the family property not even by will.[53]

Sadly, it is this simple and workable soil governance system that the Land Use Act has changed. The result is a rather chaotic and conflict laden land tenure system without any ascertainable soil governance principle. To bring out the effects of the Act, it is imperative to identify some unwholesome changes inherent in it and how the contradictions it introduced have been harvested by the ethno-political elite of the country for personal aggrandisement.

For example, this has distorted governance through avoidable herdsmen/farmers conflicts or ethno-religious tensions/conflicts that now threaten the peace and unity of the Federal Republic of Nigeria with unintended consequences.

At first blush, section 1 of the Land Use Act provides that all land in the country is to be held by the state governors in trust for the people. It has made the governors to step into the shoes of the heads of communities or heads of families, i.e. the governor of each state has to approve every land contract in his state. In other words, his consent must be obtained before any transfer or alienation of land or institutionalise trusteeship. Obviously, the general provisions of the Act have abolished, to a large extent or entirely, the customary land governance system. The first point that needs to be noted is that the prevailing statutory right of occupancy has replaced the absolute fee simple (free of charge) ownership that obtained under the customary governance system. Furthermore, sections 34 and 36 have replaced the family head with the governor and the local government council respectively. In addition, sections 24 and 25 prohibit partitioning of land into smaller portions except with the consent of the governor and the payment of highly exorbitant fees and stamp duties to the governor or the local government. More so, section 35 appears to have placed an absolute bar on the peoples' right to transfer land. Also important is the fact that sections 28 and 5 subsection 2 have made fluid the capacity to revoke

---

[50]Note, however, that this customary principle of the head of family in Southern Nigeria is similar to some other African societies. For a good comparative study see Bentsi-Enchill (1964), also see Sarbah (1968). According to him: "The village community is a corporate body, of which the members are families, or family groups, residing in the several households, and including the joint as well as patriarchal families.".

[51]Amodu v. Secretary of Southern Nigeria (1921) A.C. 399–404.

[52]The dictum of per Foster-Sutton, P. in Fynn V. Gardiner (1953) 14 W.A.C.A. P. 260. Cf. this view with the flexible position expressed by Per J Robinson in Archibong V. Archibong (1947) 18 N.L.R. 117.

[53]Ogun V. Mefun (1931) 1 N.L.R. 82. See Abeje v. Ogundairo the Nigerian Supreme Court Case (unreported SC.80/1968 dated 13-02-1970).

existing rights of the family or the community because the Act's provision refers to a "person". Unfortunately, the Act does not define the meaning of a person to include or exclude community or family. As regards the position of the local government council, section 50 defines a customary right of occupancy granted by a local government under the current Act as:

> The right of a person lawfully using or occupying land in accordance with customary law and includes a customary right of occupancy granted by a local government under this Act.[54]

Section 6 permits the above to be granted by the local government. Also, in section 36, the Act purports to convert rights held in land in non-urban areas to customary rights of occupancy. This provision is clearly inconsistent and contradictory and leads to unclear results; to say that a right is held and enjoyed in accordance with customary law and yet permits such right to be granted by a local government which is a third tier of government created by the Constitution of Nigeria.[55] The same view was held by Lord Lugard in his *Dual Mandate* as a colonial officer. According to him, land in the then Northern Region of Nigeria could not properly be described as 'native land' where the governors had the right to demand rents, to nullify all alienation without his approval and, above all, to revoke the right at will.

A critical look at events since the Act became law confirms the view that the issue of land dispute has been aggravated into more protracted and 'deadly' communal and ethnic conflicts between traditional overlords and perceived customary tenants. The family or communities still sue for a declaration of title to land.

In addition, the land conflicts have inundated the courts occasioning delays and huge financial resources to the extent that self-help is increasingly becoming the preferred option in asserting ownership rights. In some cases, a revocation order made by governors often cannot be implemented owing to the resistance by erstwhile traditional/customary land owners. Indeed, it is often difficult to carry out surveys which must form the basis of revocation cases due to resistance.

Two commonly cited judicial authorities following the promulgation of the Land Use Act decided by the Lagos State High Court disclose the nature of the problem. In *Animashaun v. sufiami*, Okuribido J. held that the defendants as customary tenants having alienated the land without the consent of the overlord forfeited their customary tenancy and the overlord was held to be entitled to resume possession of the vast area of land in dispute. This was in spite of the land Use Act. In another easily referenced case, Cole J., after hearing all the evidence and submission of the learned counsels to the parties in *Davies V. Ilo*, granted an application for a declaration of title to another vast area of land in Ikeja, having dismissed the allegation by the defendants that the plaintiffs were customary tenants. The judge concludes thus:

---

[54] *Customary right of occupancy* is defined under section 50(1) as "the right of a person or community lawfully using or occupying land in accordance with customary law and includes a customary right of occupancy granted by a Local Government under this Decree." *Statutory right of occupancy* is also defined under the same section as "a right of occupancy granted by the Military Governor under this Decree."

[55] The Constitution of Nigeria (1979).

The plaintiffs' claim, jointly and severally, against the defendants is for a declaration of title to the land in dispute. The plaintiffs are claiming as fee-simple owners of the land in dispute by virtue of the consent judgement Exhibit P4. I am satisfied on the evidence, both oral and documentary, that the plaintiffs have established their claim as fee-simple owners to the land in dispute. In exercise of the powers conferred by S.40 of the Land Use Act of 1978, I hereby declare that the plaintiffs are entitled to a statutory right of occupancy in respect to the land in dispute, verged brown on Plan No. JA/30/73, drawn by J.A. Adegboye, Licensed Surveyor, and marked as Exhibit P5 in the proceedings.

Although the Act purports to vest all the land on the state with the governor holding same in trust for the people, it is clear from the above cases that the pattern of land dispute remains unaltered. More so, the number of feuding communities and ethnic groups has increased tremendously. There is now, literally, 'armed soil governance wars' among Nigerian people[56] in almost all the 36 states of the federation.

The problem created by sections 34 and 36 already referred to above, is more pronounced in the relationship of the overlord and the customary tenant. Both according to these sections have vested interest in the land which is subject to customary tenancy. This position is further strengthened by the rule that neither of them could alienate the land without the consent of the governor of the state.

These provisions which simply provide that land is to continue to be held by the person in whom it was vested, are not only ambiguous in the sense that they do not tell us which of these persons under the customary law shall be entitled to the right of occupancy of the land,[57] but are also dangerous since they have led to a lot of misconceptions and resultant warfare. Of course, it should be expected that the overlord would continue to insist on the payment of tribute by the customary tenant, and as we have seen from *Animashaun v. Sufiami*,[58] refusal to pay tribute would amount to a denial of the overlord's title which will incur forfeiture[59] because the customary law defines a "landlord as the overlord while the tribute-paying person remains the servant or tenant".[60] It appears the Act has not affected this position

---

[56]The people of Modakeke and Ife are another example showing the attitude of the people to the land despite the Land Use Act. The Modakeke perhaps influenced by the genial intendment of the Act stopped the payment of Ishakole (tribute) to the Ife people. The Ijaws, Itsekiri and Urhobos, Aguleri Land Crisis, Tiv – Jukum, etc.

[57]Okuojenor v. Sagay [1958] W.R.L.R. 70; Ochoma v. Unosi (1960) 4 E.N.L.R. 107 and Eyamba v. Holmes (1924) 5 N.L.R. 83.

[58]Animashaun v. Sufiami (Unreported) Suit No. LD/1977 Judgement (10-06-1980).

[59]Animashaun v. Sufiami (Unreported) Suit No. LD/1977 Judgement (10-06-1980).

[60]This quotation from the Sunday Tribune (February 9th, 1986) underscores this fact. It reads: "Traditional rulers who had before the birth of the Decree been deemed custodians of land in their domain felt jilted, spited and trimmed by the Decree. During a courtesy call on ex-President Shehu Shagari (9-09-1982) six traditional rulers registered their opposition to the decree. ... In a memorandum presented to the ex-President, the traditional rulers stated that 'the traditional concept of land should be held in trust by the living for the dead as the unborn. Thus the trustees are variously the Oba or the family or even at times, the individual.' The traditional rulers noted that the Land Use Act was unnecessary as States of the Federation could legally and compulsorily acquire land for overriding public purposes with the payment of compensation."

significantly in practice. No doubt, these inherent contradictions between the Land Use Act and the customary land governance system have not addressed the yearnings of the Nigerian people for a participatory and robust soil governance and tenure system obtainable in their customs and traditions that simply means landlord and tenant.

The above argument is the reason why some scholars are compelled to insinuate that the Land Use Act was brought about as a result of ethnic and power interest in Nigeria and not to promote soil governance in every ramification. Accordingly, Agbosu observes that:

> The Act is a product of the inherent contradictions of the colonial and neo-colonial dependent, pseudo-capitalist economic structures established in Nigeria since colonial times. By the 1970s, these contradictions became so serious that they threatened to become a clog on the growth of the capitalist economy. If such contradictions had been allowed to reach a nodal point, conditions for the self-negation of the existing socio-economic and legal order would have ensued. Furthermore, the legislature, it would seem, narrowly identified the problem with private ownership of lands from its own class perspective, that is, without a scientific conception of the problems in terms of ownership in the theory of social relations. A scientific conception would have revealed the essence of the difficulties as relating not merely to the procedural aspects of private ownership of the lands, such as certainty of title, registration of title, etc. but concerning the institution of private ownership as an economic and legal category around which the exploitation of man by man is organized in class-divided societies. Such a scientific perception of the problems would have demanded a lasting solution that not only abolished private ownership rights in land but also abolished private ownership of other means of production.

> The socialisation of all means of production would have amounted to a holistic approach to the solution of the problems in the interest of the nation as a whole.[61]

The above instructive quote suggests that there have been consistent and deliberate attempts from the colonial era to manipulate ownership and access to natural resources especially through soil or land governance in Nigeria. The elite merely took advantage of already established contradictions after independence.

5   Soil Protection Provisions in Nigeria

Soil protection can be viewed from a biodiversity approach.[62] Accordingly, the drafting of law and policy on soil can be anchored on an ecosystem approach. This implies that sufficient legal mechanisms in favour of biological diversity perspective are used in framing soil protection instruments. This approach incorporates in itself the environment and the landscape. For sustainable use of soils conditions to be developed, "governance programs must be integrated from the local to global level, across a range of sectors, and over a substantial time frame to

---

[61]Agbosu (1988).
[62]Boer and Hannam (2015).

enable effective soil policy making."[63] Further, "Soil protection and rehabilitation policies need to be based on a human right framework, principally emphasising land rights for marginal and vulnerable groups in society".[64] This implies that groups such as women and minorities should be considered in the decision-making processes. This seems to be the challenge Nigeria as a nation must come to terms with when developing new and comprehensive instruments on soil protection.

Contemptuously, the Nigerian Federal Government has also not availed itself with the few sustainable, development based, capacity building regional and international collaborations with development partners, the private sector and individuals to address soil governance. Neither does it considers enforcing coherent policies that encourage practices and methodologies that can regulate the usage of soil or land as a natural resource. Collaborations of this nature play a significant role on the avoidance of conflicts between land users. They promote sustainable land management that ultimately increase and sustain soil integrity in terms of fertility and output in relation to agricultural yield ascertainable as land degradation neutrality, as conceived in the SDGs.

There are, however, limited and uncoordinated principles that pertain to soils, particularly, in the area of evaluation of contaminated sites. In other words, these are only contained in part, in other legislations like the National Environmental Standards Enforcement Agency (NESREA), or Urban Planning and Building Laws, etc. but are rarely implemented or enforced.

The current Nigeria federal law that governs soil and land tenure is the Land Use Act. Unfortunately, this legislation does not address the issues associated with sustainable soil or land development highlighted above. It also cannot be described as a law, policy, strategy, or any other conceivable process of decision-making that is coherent with equitable use of soil. By every standard of analysis, the current Act was not developed with the active participation of the citizenry. On the contrary, it was superimposed on the nation by the then military government. Above all, it does not include restoration or remediation of degraded soils. Another glaring lacuna is that it has proved to be incapable of holding violators accountable. After about 40 years in force, the Act has not been able to guarantee healthy and productive soils for a food-secured country, as well as supporting other essential ecosystem services.

Despite these shortcomings and the ensuing conflict, the political leadership at the federal level of governance does not see the urgency or need for a general object specific soil governance regime. This implies a legislation that contains provisions that effectively deal with soil protection and prevent further deterioration of soil quality. The provisions of such an instrument must preserve and promote the symbiotic, yet multiple soil functions. Above all, they must be capable of maintaining soil fertility with scientifically proven restoration system of damaged soils; ensure prudent remediation capable of restoring functionality, as well as repealing all obnoxious land and mineral legislations that do not make the polluters

---

[63]Boer and Hannam (2015).
[64]Boer and Hannam (2015).

pay for the despoliation of polluted soils. Unfortunately, there is hardly a ray of hope for victims of soil-related violence in Nigeria and even at the international level presently, because the three main international treaties which contain relevant provisions on soil protection do not deal with violence.

These instruments include the UN Convention to Combat Desertification of 1994, the Convention on Biological Diversity of 1992 and the Climate Framework Convention of 1992, and the Paris Agreement on Climate Change. The fact is that a careful perusal of these international instruments (laws) reveals that the main aims of the Desertification Convention are to combat desertification and to mitigate the effects of drought. Whereas the Convention on Biological Diversity focuses on the preservation and sustainable use of biological diversity, including terrestrial ecosystems, the Climate Change Framework Convention contains agreements on mitigation and adaptation measures, which include greenhouse gases sinks and reservoirs. Although there is a regional, international law that deals with limited soil or land governance, this African regional agreement is, to say the least, 'a toothless bull dog' without a functional secretariat due to lack of funds.

6 International Framework on Soil Governance

The significance of soil to humanity cannot be over emphasised. The Revised World Soil Charter of 2014 describes soil as "a key enabling resource, central to the creation of a host goods and services integral to ecosystems and human well-being."[65] In spite of this indispensable value, soil has not received the desired attention that guarantees its protection.

This neglect has cost the globe so much of productive soil. It is against this backdrop that states, organisations and individuals have begun to meet at different levels to proffer solutions to the challenge. The idea is to provide a legal framework that can guide against the improper use of soil as well as restoring sustaining it for future generations.

2015 was a turning point for soil governance all over the world. The first international year of soil declared by UN was held in 2015 and 5th December was declared World Soil Day that same year. Throughout the year, a lot of activities connected to soil took place across the globe. In Germany, the International Year of Soil was launched alongside two other conferences. Also, parties to the UNCCD at the 12th Conference in October 2015 agreed in Ankara, Turkey to include the objective of land degradation neutrality in its mandate. Although, "Attempt to put in place an overarching regulations at the European level (European Soil Framework Directive of 2016) failed due to a blocking minority of some European member states that considered soil to be a domestic issue,"[66] the 7th Environmental Action

---

[65]Boer and Hannam (2015).

[66]Ginzky et al. (2016).

Programme tackles soil and land issues. In particular, in Germany, there exists the Soil Protection Act which concentrates more "on the restoration of existing brown fields".[67] Nonetheless, a lot of other land and soil-related regulations are scattered in other various acts and ordinances.

Similarly, a conference on soil governance was held in Brasilia in March 2015. The Brazilian Federal Court of Accounts (TCU) is empowered by the constitution to conduct a performance audit on non-urban soil and land governance. The audit was able to examine the risk facing the natural heritage and identified the damage it has suffered. TCU has touched five of the SDGs. Goal 15 of the SDGs recognised the crisis facing soils all over the world when it stated that ". . .protect, restore and promote the sustainable use of terrestrial ecosystems, sustainably manage the forest, combat desertification, halt and reverse land degradation, halt biodiversity loss."[68] It was also proved that Brazil does not have a specific soil policy.

From the African point of view, the continent needs "a synergistic, pragmatic and adaptive approach if we are to halt, reverse and address the issue of soil degradation."[69] In Namibia, the Ministry of Agriculture, Water and Forestry (MAWF) is saddled with the responsibility of soil management, promotion and development of sustainable soil management practices in agriculture, water and forestry sectors through appropriate policy and legal instruments. The Namibian policies encourage inclusiveness which has resulted in increased food production. Despite some challenges in the process of land redistribution, there is a comprehensive framework for the reform of land tenure, acquisition and ownership.[70] From the foregoing, Namibia seems to be among the nations that are responsive to the promotion of sustainable land use.

Although there are pockets of provisions relating to soil embedded in some national and international instruments, there is an ongoing advocacy on a fair deal for soil sustainability similar to the biodiversity loss, climate change, and food security legal frameworks.[71]

Apparently, the level of consideration accorded these phenomena is heartwarming, despite the fact that soil protection is an essential factor to the success of such considerations due to the bond it shares with the other three. The 2015 Soil Atlas asserts that all the more than 200 internal treaties, agreements and protocols neglect soil conservation and fail to define specific targets. This is not an encouraging development if the world truly desires to overcome resultant effects of soil degradation. The disregard for soil protection is the cause of many of the world's current predicaments. The situation in Nigeria mentioned in the previous section is just the tip of the iceberg. There are issues of traditional notions of land ownership, particularly whether native laws and customs 'abhor alienation' of land, widespread

---

[67]Ginzky et al. (2016).

[68]Ginzky et al. (2016).

[69]Ginzky et al. (2016).

[70]Ginzky et al. (2016).

[71]Boer and Hannam (2015).

food insecurity, and many other key issues. It is in a bid to salvage or reverse this ugly trend that advocates of soil degradation neutrality call for legal and enforceable soil governance that will focus basically on battling the soil degradation scourge and to push for the preservation and sustainable use of soil. This requires a synergy among the major stakeholders (national and international). In other words, safeguarding the soil to improve human welfare is now a moral obligation. In this regard it has been opined that a treaty on the conservation and sustainable use of soils should be developed.[72] In developing the said instrument, there is need to pay attention to the United Nations Development Goals on the achievement of zero net degradation, the revision of the UN Special Rapporteur on the right to food, etc.

7 Recommendation

In Nigeria, experiences from conflict embroiled regions have heightened the anxiety and concern of the academia over the rising ethno-religious tension in the country due largely to Fulani Muslims and Christian minority ethnic farmers who struggle over access to soil or land resources. A misplaced or mischaracterised land or soil governance legislation should not be encouraged to swiftly wipe out an age long culture of ethnic minority farmers of southern Nigerian, in order to achieve economic and social goals of majority ethnic groups. Consequently, the following vexed issues need to be addressed urgently before Nigeria slides into armed conflict or disintegration over avoidable soil governance and land tenure conflicts. This is more so because Nigeria can draw lessons from other environments or natural resources induced conflicts resolution mechanisms that have worked elsewhere as sustainable legal and policy frameworks capable of addressing the issues. In the light of the above, the following recommendations are necessary and ought to be taken seriously:

• The current soil or land governance structures and ownership principles that do not recognize African perspectives on land ownership, family, community or royal heads, traditional sustainable soils or land management, entrenched in the Federal Constitution appear to be over centralized and culturally insensitive to balanced access to land or soil resources in a multicultural society like Nigeria.
• There is the urgent need to provide an enabling framework for holistic soil governance legislation with the active participation of all stakeholders.
• A mutually beneficial land tenure system with inbuilt peaceful coexistence incentives that reward peaceful settlement of soil-related disputes needs to be put in place.
• Nigerians in authority should avoid sectionalism while dealing with matters of conflict, access to resources, especially, on land ownership and ensure that current

---

[72]Boer and Hannam (2015).

obnoxious Land Use Act and minerals' legislations are repealed; the various contradictions or inconsistencies that deny minorities their age-long ownership rights are expunged from current legislations including the Constitution.

- While designing any new legal framework on soil or land governance, attention should be paid to the customary and traditional model of land ownership across the country.
- An impartial enforcement or implementation of detailed soil governance standards in all tiers of governance, and standards in other regions on uncertain climatic change and overlapping systems of resource exploitation and access to land-based natural resources is urgently required in Nigeria.
- A sustainable and responsible land and soil agenda is required to achieve the SDGs in Nigeria.
- Nigeria needs a comprehensive transformative land or soil object specific legislation that takes into consideration adaptation mechanism to effects of climate change and power imbalances.
- The sustainable use of natural resources has to be integrated into a participatory EIA processes and accountability mechanisms based on expected compliance outcomes to achieve the SDGs which are essentially natural resources based on soils or land.
- The Nigerian Federal Government must ensure neutrality of all its law enforcement agencies, including the military, engaged in peace keeping or law and order enforcement on the Fulani Muslim Herdsmen and the indigenous Christian Ethnic Minority Farmers crises.
- Appropriate accountability mechanisms within the post-2015 Development Agenda should be encouraged to address the challenges.
- To contribute to the successful implementation of the post-2015 Development Agenda, the soil governance research pattern needs to change too. Land and soil governance research must be more systematically conducted in an interdisciplinary manner and must partner with ongoing law reform and transformation processes.
- Tension prone regions or states should be identified and mapped out, and all feuding parties (Fulani herdsmen, Christian farmers, howsoever described) must be disarmed.
- Immediate provision of security to protect lives and property in the conflict areas, particularly, the Christian farmers' territories and the peaceful return of Internally Displaced Persons (IDP) is necessary.
- The establishment and institutionalization of local peace building for negotiation and conflict resolution must be developed.
- Science based transformative research sensitive to the multi-ethnic cultures of the Nigerian society needs to be added into the ministerial lines of duties as a basis for routine engagement in governance design and policy implementation.

8 Conclusion

This chapter is written in the context of Nigeria's emerging fiasco in the sphere of 'natural resource governance', or conflict laden soil governance legal and policy framework. This, however, is against the ground swell of documented warnings by several scholars who had pointed to the blanket adoption of Eurocentric alien laws, policies and values without adaptation to local circumstances. The end to land or soil governance-related conflicts in Nigeria seems not quite close, except the federal government takes bold and convincing steps to tackle the root causes head on. Regrettably, the leaders of the country at different levels, especially the elite, seem to be comfortable with the failed soil/land governance system. Invariably, the government needs to explore all available avenues from both local and international perspectives to engage all stakeholders to arrive at a well-articulated and workable legal framework that is relatively acceptable to the various groups and segments of the Nigerian federation to achieve a positive result.

Although the soil or land governance issue is a global phenomenon, every nation has its peculiarities. In Nigeria, it is made more complicated with the introduction of ethno-religious and political 'face' into it. More so, lack of political will and deliberate mischaracterization of the issue by the federal government coupled with 'perceived incapacity' to tackle the root causes of the problem, drastically, makes matters worse. This portends avoidable tensions and more crises for the country as more pressure is placed on the limited land and soil resources through the activities of non-indigenous groups.

References

Adigun O, Omotola JA (1982) The Land Use Act – Report of a National Workshop

Agbosu L (1988) The Land Use Act and the state of Nigerian land law. J Afr Law 32(1):1–43

Allaby M, Park C (eds) (2017) A dictionary of environment and conservation, 3rd edn. Oxford University Press

Bentsi-Enchill K (1964) Ghana land law, an exposition, analysis and critique. African University Press, Lagos

Boer B, Hannam I (2015) Developing a global soil regime. Int J Rural Law Policy Soil Gov, Special Edition 1

Coker GBA (1966) Family property among the Yorubas (No. 14). Sweet and Maxwell

Daily Post (January 16th, 2018a) Senator Adamu Aliero, Dino Melaye differ on creation of cattle colony. Available at: https://dailypost.ng/2018/01/16/senator-adamu-aliero-dino-melaye-differ-creation-cattle-colony/

Daily Post (March 26th, 2018b) Sagay backs Danjuma's self-defence advice, says law allows Nigerians to defend themselves. Available at: https://www.thecable.ng

Daily Post (April 24th, 2018c) Catholic Church confirms killing of Rev Father in Benue. http://www.google.com/amp/dailypost.ng/2018/04/24/catholic-church-confirms-killing-rev-fathers-benue/ampshare=http://dailypost.ng/2018/04/24/catholic-church

Duggan CS (2009) Nigeria: opportunity in crisis? Harvard Business School

Elias TO (1971) Ilorin land and the land tenure law 1962. Niger Law J 5(1971):159

Garner BA (ed) (2015) Black's law dictionary, 10th edn. Thomson Reuters West (May 18th, 2015)

Ginzky H, Heuser I, Quin T et al (eds) (2016) International Yearbook of Soil Law and Policy 2016. Springer International Publishing 2017

Junge B, Mabit L, Dercon G, Walling DE, Abaidoo R, Chikoye D, Stahr K (2010) First use of the 137Cs technique in Nigeria for estimating medium-term soil redistribution rates on cultivated farmland. Soil Tillage Res 110:211–220

National Population Commission Nigeria, NPC (1998) 1991 Population Census of the Federal Republic of Nigeria: analytical report at the national level. Lagos, Nigeria

Orubebe BB (2017) Comparative environmental governance, law and policy: an analysis of judicial techniques in India and Nigeria. Comp Law Rev 23(2017):50–81. Available at: https://apcz. umk.pl/czasopisma/index.php/CLR/article/view/CLR.2017.002

Osaghae EE (1998) Crippled giant: Nigeria since independence. Indiana University Press

PM News (January 30th, 2018) Cattle colony: before president Buhari lures us into deadly trap. Available at https://www.pmnewsnigeria.com/2018/01/30/cattle-colony-buhari-lures-us-deadly-trap/

PM News Nigeria (February 12th, 2018) Fulani herdsmen burn falayes farm. Retrieved from: https://www.pmnewsnigeria.com/2018/01/21/Fulani-herdsmen-burn-falayes-farm

Ponge JF (2015) The soil as an ecosystem. Biol Fertil Soils 51(6):645–648. Available at: https:// www.researchgate.net/publication/276090499_The_soil_as_an_ecosystem

Premium Times Nigeria (April 12th, 2018a) Buhari blames Gaddafi for killings across Nigeria. Available at: https://www.premiumtimesng.com/news/top-news/264764-buhari-blames-gaddafi-for-killings-across-nigeria.html

Premium Times Nigeria (April 30th, 2018b) Nigeria sliding from ethnic cleansing to genocide-Soyinka. Available at: https://www.premiumtimesng.com/news/headlines/266624-nigeria-slid ing-from-ethnic-cleansing-to-genocide-soyinka.html

Pulse Newspapers Nigeria (July 4th, 2018) Police arrest suspected herdsman with AK-47. Available at: https://www.pulse.ng/news/local/in-enugu-police-arrest-suspected-herdsman-with-ak-47/ 54vw1hw

Punch Newspapers Nigeria (January 15th, 2018a) Accommodate your countrymen, Buhari begs Benue leaders. Available at: https://punchng.com/accommodate-your-countrymen-buhari-begs-benue-leaders/

Punch Newspapers Nigeria (May 2nd, 2018b) Stop wanton killing of Benue people. Available at: https://punchng.com/stop-wanton-killing-of-benue-people/

Quartz Africa (May 1st, 2018) President Buhari's meeting with Donald Trump was all about staying on message. Retrieved from: http://qz.com/1266514/president-buharis-meeting-with-donald-trump-was-all-about-staying-on-message

Sarbah JM (1968) Fanti customary laws, 3rd edn. Africana Modern Library No. 5, London

Situma DFP (2003) Legislative & institutional framework for Community Based Natural Resource Management in Kenya. Univ Nairobi Law J 1:55–60

The Cable (April 27th, 2018) Tension as angry youth attack Hausa community in Benue, 'kill 8'. Available at: https://www.thecable.ng/tension-angry-youth-attack-hausa-community-benue-kills-8

The Guardian Nigeria (April 13th, 2018) Uproar in Senate over Buhari blaming Gadaffi for killings. Available at: https://guardian.ng/news/uproar-in-senate-over-buhari-blaming-gadaffi-for-killings/

Udo RK (1970) Geographical regions of Nigeria. University of California Press

United Nations Convention on Combating Desertification (UNCCD), Conference of the Parties. Twelfth Session, COP 15 (2018) Integration of the Sustainable Development Goals and targets into the implementation of the United Nations Convention to Combat Desertification and the report of the Intergovernmental Working Group on Land Degradation Neutrality- ICCD/COP (12)/4). Available at: https://www.unccd.int/sites/default/files/sessions/documents/ICCD_COP12_4/4eng.pdf

United Nations Environment Programme, UNEP (2016) Mid-Term Strategy 2018–2021. Environ-
    mental Governance, pp 33–36. Available at: https://www.unenvironment.org/annualreport/
    2018/index.php#ch-02
Vanguard (February 28th, 2018a) Federal government gives reason for proposing cattle colonies.
    Available at https://www.vanguardngr.com/2018/01/fg-gives-reason-proposing-cattle-colonies/
Vanguard (March 24th, 2018b) Defend yourselves or you will all die, TY Danjuma tells Nigerians.
    Available   at: https://www.vanguardngr.com/2018/03/defend-will-die-ty-danjuma-tells-
    nigerians/
Vanguard (April 30th, 2018c) Nigeria: rowdy session as PDP senator calls Buhari incompetent.
    Available at: https://allafrica.com/stories/201804130146.html

Soil Legislation in Australia

Ian Hannam

1 Introduction

Soil conservation legislation was first introduced in Australia over 80 years ago. It was introduced primarily to control land degradation. The term 'degradation' is used to describe changes that are additional to those occurring naturally and carries with it the notion of change that is undesirable and brought about by humans. Land degradation encompasses many types of soil degradation and refers to chemical and biophysical changes in the land that reduce both its quantity and quality. In Australia, these changes are linked to a reduction in the productive capacity of land and its economic value.

The condition of land or 'land health' invokes the concept of ecosystems—the interactions and connections between the living and non-living components of the environment. In degraded land, where ecosystems have been changed, the altered ecosystems continue to function but have a reduced capacity to supply the goods and services we are seeking, for example, food, habitat for threatened species and landscape amenities.

In addition to providing for the physical needs of increasing populations, in Australia, the land has a spiritual and cultural significance for many but in particular Indigenous people. The changes in the condition of land following settlement by Europeans in 1788 have diminished these values in many parts of Australia, although some older cultural sites have only been revealed following disturbance by the new settlers. In Australia, about two thirds of agricultural land is degraded. The major types of land degradation are soil erosion, soil salinity, soil acidity, soil

I. Hannam (✉)
University of New England, Australian Centre for Agriculture and Law, Armidale, NSW, Australia

© Springer Nature Switzerland AG 2020
H. Yahyah et al. (eds.), *Legal Instruments for Sustainable Soil Management in Africa*, International Yearbook of Soil Law and Policy,
https://doi.org/10.1007/978-3-030-36004-7_10

contamination, nutrient loss and soil structure decline.[1] Soil conservation legislation was one of the first areas of natural resource law and remained prominent until the 1980s and has been introduced to deal with the different forms of land degradation.

Australia is a federation of a national government, six states and two territory governments (i.e. nine jurisdictions).[2] Legislatively, soil conservation is a state responsibility and, although the Australian Constitution 1900[3] (hereafter the "Constitution") does not provide specifically for the natural environment, this has not prevented the Commonwealth from taking a comprehensive and active leadership in soil conservation, from a strategic perspective and providing financial resources.[4] At the state level, seven of the eight jurisdictions had some form of soil conservation law until the 1980s. By 2018 only five states retained a specialist soil conservation law. In the 20 years between 1980 and 2000, the activity of soil conservation was increasingly viewed as a major ecological issue in Australia and integrated within comprehensive environmental law systems. In this regard, the remaining specialist soil conservation laws play a subordinate role to the integrated natural resources laws in environmental assessment, planning and management of soil resources.[5] Using the standard Australian definition of 'soil conservation' as a guide as to which consolidated laws relate to that definition indicates that there are around 200 individual laws in Australia that play a prominent role in the implementation of soil conservation.[6] These laws are categorised into three separate categories according to the level they contribute to achieving the soil conservation objective.[7]

---

[1] Australia State of the Environment 1996 and 2006. These reports describe the various forms of land degradation in Australia, their extent and, where data is available, trends.

[2] In this chapter a reference to "states" means the six state and two territory governments of the Australian mainland. The difference between Australian states and territories lies in the governing powers of the states and territories. Australia is a huge country and a continent in itself. It is referred to as a Commonwealth of Australia being a union of 6 states and the two territories (in total there are 10 Australian territories). This division between states and territories has been done for administrative convenience. Australian states came into existence even before the federal government came into power, and these states have their powers protected in Australian constitution. Territories are under the direct control of the federal government, and parliament has powers to legislate for territories while it cannot legislate for states.

[3] The Commonwealth of Australia Constitution Act 1900.

[4] The Commonwealth of Australia is the official name for Australia, which is a nation occupying the whole of the Australian continent; Aboriginal tribes are thought to have migrated from southeastern Asia over 20,000 years ago; first Europeans were British convicts sent there as a penal colony. The Federation of Australia was the process by which the six separate British self-governing colonies of Queensland, New South Wales, Victoria, Tasmania, South Australia, and Western Australia agreed to unite and form the Commonwealth of Australia, establishing a system of federalism in Australia.

[5] Hannam (2006).

[6] The consolidated acts in the nine jurisdictions in the Australian Legal Institute database "austlii" were interrogated.

[7] It should be noted that various laws under the three categories play a role in controlling soil degradation which occurs from urbanisation and industrial uses.

The Category 1 specialist soil conservation laws contain a wide range of provisions to implement practical soil conservation. They have a range of duties, powers and functions that enable the formation of committees, advisory groups, to undertake research, experimentation, education, financial arrangements, land use planning, land management, and legal enforcement. However, due to the growth of the environmental movement in Australia over the past thirty years, the specialist "single issue" natural resource laws (e.g. soil conservation, water, forestry) have either been superseded by comprehensive integrated resource laws and only used to guide practical conservation, or repealed and their main functions incorporated within a comprehensive integrated resource law. In this regard, it is the Category 2 system of legislation in Australia that now performs the major legislative responsibilities for soil conservation, as part of their broad role in environmental management. This situation is a direct outcome of the changing role and perception of environmental law in Australia, which has largely come about through the increasing intervention of the community and public participation in the management of the environment.[8] This approach is more effective in being able to manage the total environment, and government authorities in all jurisdictions have been reorganised so that no Australian jurisdiction now has a specialist soil conservation authority. Soil conservation activities have become the responsibility of multidisciplinary organisations with integrated resource management responsibilities. In this context, decisions about the utilisation of soil resources are made in an ecosystem context. The Category 3 laws provide an important role in supporting the soil conservation objective in the areas of bushfire management, mining, forestry, agricultural tenancy and primary industry funding, for example.

One of the main characteristics in the management of natural resources in Australia is the level to which international environmental strategies are used in the development of Australia's grand strategy for environmental management, including their use in framing natural resources laws. This approach has led to a more equitable allocation of financial and human resources in the control of soil degradation and biodiversity management under major national programs, e.g. the National Soil Conservation Program of the 1980s and the Natural Heritage Trust introduced in 1997. It has also influenced the formation of major environmental laws such as the Water Act 2007 (Murray-Darling Basin) and the National Environment Protection Council Act 1994, which are "mirrored" at the state level. The Australian situation clearly shows the changing relationships between the national and state level of government in the management of natural resources. Although the Australian Constitution makes no specific provision for environmental issues, and the Commonwealth government had no direct role in soil conservation for many years, by necessity, as these issues became more ecologically, administratively, financially and economically complex, there has been a need to change national-state relations in environmental management. From a legislative perspective, the law must evolve and adapt to meet the above types of changes. In the Australian case, the

---

[8]Bates (2016).

main driving force has been ever-increasing demands from the community for improved environmental management, including more comprehensive and capable-but flexible law-making processes. In this regard, clearer roles and lines of responsibility have developed between the national and state jurisdictions, particularly in the approach to the development of natural resources law. This has led to a more innovative and well-defined approach to natural resources law, and the current approach to manage soil resources is a good example of this.[9]

Laws in the three categories are important for achieving the United Nation's objective for a land degradation neutral world. In this regard, paragraph 205 of *The Future We Want*, contextualises the issues of desertification, degradation and drought.[10] It recognised 'the economic and social significance of good land management, including soil, particularly its contribution to economic growth, biodiversity, sustainable agriculture and food security, eradicating poverty, women's empowerment, addressing climate change and improving water availability'. It also urged that 'desertification, land degradation and drought are challenges of a global dimension and continue to pose serious challenges to the sustainable development of all countries, in particular developing countries. Further, paragraph 206 recognises, 'the need for urgent action to reverse land degradation. In view of this, we will strive to achieve a land-degradation neutral world in the context of sustainable development.' Paragraph 207 sets out a broad framework for national, regional and international action in order to monitor land degradation and to restore degraded lands in arid, semi-arid and dry sub-humid areas.

2   Development of Soil Conservation in Australia

In the Australian *Glossary of Terms Used in Soil Conservation*, "soil conservation" is defined as the "prevention, mitigation or control of soil erosion and degradation through the application to land of cultural, vegetative, structural and land management measures, either singly or in combination, which enables stability and productivity to be maintained for future generations".[11] In this definition, the reference to "structural and land management measures" includes the methods and approaches used to prevent soil erosion and restore land.[12] These activities are reflected in the

---

[9]Bates (2016).

[10]The Future We Want, UN Doc A/66/L 56.

[11]Houghton and Charman (1986), p. 116.

[12]E.g. see reference in Preamble to the New South Wales Soil Conservation Act 1938 as "an Act to make provision for the conservation of soil resources and farm water resources and for the mitigation of erosion"; under Section 4 of the Western Australian Soil and Land Conservation Act 1945 "soil conservation" means the application to land of cultural, vegetation and land management measures, either singly or in combination, to attain and maintain an appropriate level of land use and stability of that land in perpetuity and includes the use of measures to prevent or mitigate the effects of land degradation; under Section 3 of the Northern Territory Soil

procedures found within the various soil conservation laws and can be directed at land degradation processes associated with agriculture, urbanisation, and industrial land uses.[13]

For over 80 years, Australia has used a variety of technical, institutional, legislative, and strategic tools to achieve soil conservation. During the first 50 years, the soil conservation effort remained relatively unchanged.[14] In the past 30 years, however, the political, policy and legislative aspects of the soil conservation discipline have changed significantly. By the mid-1900s most states had introduced soil conservation legislation and established some form of soil conservation institution, or authority.[15] Some of these institutions developed highly integrated approaches to soil conservation and became world-renowned, e.g. the Soil Conservation Service of New South Wales.[16] Traditionally, soil conservation legislative responsibilities focused on the control and mitigation of water and wind erosion in agricultural and pastoral areas, in some cases from a catchment or river basin perspective.[17] Close interaction and cooperation with farmers and pastoralists were a hallmark of these institutions. The Australian federal government maintained a coordinating role with the states and the Australian Standing Committee on Soil Conservation, formed in the 1950s had representatives from all states and the federal government. However, implementation remained, primarily, a state responsibility. A significant turning point for soil conservation in Australia was in the late 1970s when the federal government coordinated a national evaluation of land degradation and soil conservation strategies and policies—"*A Basis for Soil Conservation Policy in Australia*" (hereafter the "Collaborative Study").[18]

Conservation and Land Utilization Act 1980 "soil conservation treatment" means structural or agronomic work for the purpose of the conservation of soil or reclamation of land.

[13]New South Wales Soil Conservation Act 1938 Section 3, reference to "works" as (a) works necessary for the conservation of soil or the mitigation of erosion and any operations incidental thereto, or (b) works necessary for the conservation of water resources or the provision or improvement of the water supply to farming lands for domestic or stock purposes"; under Section 6 of the Queensland Soil Conservation Act 1986, "soil conservation measures" means works, land management practices, undertakings, acts, proposals, prohibitions and things designed, carried out, enforced or proposed to be carried out or enforced pursuant to this Act for the purpose of soil conservation or controlling or directing run-off water flow.

[14]Downes (1970) and Bradsen (1988).

[15]Bradsen (1988).

[16]Breckwoldt (1988).

[17]Hannam (2003), pp. 112–115.

[18]Australia (1978a) Report 1; Report 14.

3   Integrated Environmental Management

The *National Soil Conservation Program* (NSCP) was established in 1983 as the primary means to implement the major findings of the Collaborative Study and to confront Australia's serious soil degradation problem.[19] States were allocated finance by the federal government to stimulate their soil degradation control efforts, and the *National Soil Conservation Advisory Committee* (NSCAC) recommended soil conservation priorities and strategies based on assessed national needs. Significantly, around this time, the perception and understanding of soil conservation began to change. Farmers and environmentalists became united in their view that soil conservation was a major national environmental issue, a view strongly influenced by an expanding global movement for sustainable development and this was reflected in the 1989 *National Soil Conservation Strategy of Australia* (NSCSA),[20] which was developed in response to the *National Conservation Strategy for Australia*.[21] Another significant development in the 1980s was that the effectiveness of soil degradation policies was being brought under serious question, widening the debate over the environmental role of soil conservation and its potential contribution to biodiversity conservation.[22] A variety of national and state environmental strategies were developed during this period, acknowledging the contribution of soil conservation to improved water quality, landscape health, to increasing soil carbon storage as a component of climate change management, and to the conservation of Australia's biodiversity.[23]

By the early 1990s, in keeping with the national agenda of environmentalism—which was consolidated in the nation's *Intergovernmental Agreement on the Environment* (IAE)[24]—most states by this time had commenced comprehensive environmental law and policy reform. The reform occurred in the areas of water, vegetation, environmental assessment, environment protection, land planning, and pollution management and with a greater focus on integrated natural resources legislation. The traditional "soil conservation" law was not a prominent part of this reform, with only two of the eight states replacing their soil conservation laws. The eight states reacted as follows:

- Four states retained their soil conservation laws but introduced additional integrated natural resource management laws—Queensland, New South Wales, Western Australia, and Northern Territory.
- Two states abolished their soil conservation laws and introduced comprehensive integrated natural resources law that made provision for soil conservation as a

---

[19]Australia (1978b), Report 1 Chapter 4, Land degradation assessment.
[20]Australian Soil Conservation Council (1989).
[21]Australia (1983).
[22]Australia (1989).
[23]Australia (1992b).
[24]Australia (1992a).

general component of broad natural resource management—Victoria and South Australia.

- The Australian Capital Territory did not have a specific soil conservation law, as such, adopting the Soil Conservation Ordinance in 1960. Now, soil conservation provisions are part of a comprehensive environmental legislative regime.[25]
- The State of Tasmania has never had a specific form of soil conservation legal instrument, and in this state soil conservation activities have been part of the state agriculture department regime.

The current legislative system for soil conservation is such that each individual jurisdiction has a number of primary environmental laws to establish the standards, rules, policies, auditing and compliance responsibilities for natural resources management (i.e. sustainable land management, ecologically sustainable development, integrated natural resource management, biodiversity management). In general, there is one principal law that has a major coordination role and often sets the rules and standards for environmental management.[26] Under these circumstances, the "single issue" legislation for soil, water, forestry and vegetation, play a supplementary role to the responsibilities of the primary environmental laws, where they set out basic natural resource management standards for these single resource areas which are then implemented through the functions of the primary environmental law mechanisms.

By the mid-2000s, the main influence on state soil conservation efforts derived from the implementation of the national Natural Heritage Trust Program (NHT).[27] The NHT was one of the most comprehensive environmental resource management

---

[25]Includes the Environment Protection Act 1997 and the Planning and Development Act 2007.

[26]E.g. the Environment Protection and Biodiversity Conservation Act 1999 at the Commonwealth level; see Hannam and Boer (2004) for a detailed discussion on essential elements for soil legislation.

[27]The Natural Heritage Trust was set up by the Australian Government in 1997 to help restore and conserve Australia's environment and natural resources. Since then, considerable numbers of community groups and organizations have received funding for environmental and natural resource management projects. The Natural Heritage Trust ceased to operate on 30 June 2008. Its function was included in the work of the 'Caring for our Country' funding program. Caring for our Country was an initiative that offered multi-year funding to provide certainty for stakeholders. The Australian Government announced that Caring for our Country would be combined with the National Landcare Programme in 2013. The National Landcare Programme is continuing to deliver upon initiatives that were in place before 1 July 2014. The delivery of the second phase of 'Caring for our Country' (2013–2018) was through two specific streams, **Sustainable Environment** and **Sustainable Agriculture.** The Sustainable Environment stream aimed to ensure Australia's national environmental assets are conserved, resilient and healthy. The Environment Protection and Biodiversity Conservation Act 1999 focuses Australian Government interests on the protection of matters of national environmental significance—nationally and internationally important flora, fauna, ecological communities and heritage. Caring for our Country investment under the strategic objectives was guided by these legislative responsibilities. The Sustainable Agriculture stream was delivered in the context of other major government policies and initiatives including the National Food Plan, the Intergovernmental Agreement on Biosecurity, the Carbon Farming Initiative, drought policy reform and the National Volunteer Strategy.

programs undertaken in Australia. Moreover, the national Landcare program[28] and other environmental programs, operate through "self-starting" community groups to address soil, water and biodiversity problems.[29] Bilateral agreements have been negotiated between the national government and the states to implement the framework first devised under the NHT. Funds are provided for community-based activities, rather than traditional, technical-based soil conservation activities which remain the responsibility of the states where they are implemented by multifunctional and multidisciplinary state natural resource agencies. The approach adopted by each state varies, but in some states, the agencies have several natural resource-related responsibilities, including forest management, coastal management, native vegetation management, national parks and wildlife, and fisheries management, or different combinations of these.[30] This multidisciplinary and integrated approach to natural resources management in Australia provides substantial flexibility in designing approaches to soil conservation.

Because of the above factors, national-state soil conservation activities, unquestionably, have expanded and improved substantially in Australia over the past 30 years. Significantly, more human and financial resources have been transferred into the integrated, community-based natural resource management activities, in keeping with the broad environmental agenda of Australia.[31] Considerable work has been undertaken on the development and adaptation of farming systems, that, ecologically, maintain or enhance the natural resource base and many landholders have made significant changes to their farming practices. The first national survey of land degradation was undertaken in the mid-1970s, and in 1987–1988 the first systematic land degradation survey was undertaken of New South Wales which indicated that almost every part of the state was affected by one or more forms of land degradation.[32] It has been established that around 2% of Australia has soils

---

[28]The National Landcare Program is a key part of the Australian Government's commitment to protect and conserve Australia's water, soil, plants, animals and ecosystems, as well as support the productive and sustainable use of these valuable resources.

[29]The Australian Natural Resources Atlas was developed by the National Land and Water Resources Audit to provide online access to information to support natural resource management. The Atlas was managed and maintained within the Department of the Environment, Water, Heritage and the Arts. The Atlas comprises a number of tools and information on Australia's natural resources.

[30]E.g. Commonwealth—Department of Agriculture and Water Resources and Department of the Environment, Water, Heritage and the Arts; Victoria—Department of Environment, Land, Water and Planning; New South Wales—Office of Environment and Heritage, Department of Primary Industries, Local Land Services; Western Australia—Department of Environment and Conservation; Queensland—Department of Natural Resources, Mines and Energy, Department of Agriculture and Fisheries.

[31]Martin (2017), p. 34 suggests that 'any policy that depends on substantial government investment must be considered to have a limited chance of being successfully adopted. A return to the golden age of public investment in soils governance, much as it might be desired, seems unlikely for economic reasons.'

[32]Graham (1992), p. 205.

regarded as excellent quality, and 70% of Australia is comprised of soil not usable for agriculture.[33] Although soil conservation law has existed in Australia for around 80 years, many state laws lacked specific objectives and targets to reduce soil degradation and many, unlike the modern environmental laws, had restricted powers to deal with issues like soil and water salinity, addressing nutrient loss, and managing toxic soils. The early laws had several basic characteristics in common, including administrative features, powers, duties and functions, interagency relationships, cooperation and coordination, objectives, land use planning, and financial assistance provisions. The South Australian Soil Conservation and Land Care Act of 1989[34] was an exception, as it contained provisions for a land capability-based approach to land use and land degradation control. When viewed as a specific group or class of environmental law in Australia, soil conservation law, by comparison to other types of environmental law (e.g. forests, endangered species, environmental planning) does not have an ecological-based strategy. The legislation is characterized by provisions that concentrate on practical land management measures to "protect" agriculture, rather than provisions that can adequately determine the ecological constraints of natural resources and take sustainable land use action. The major attitudinal change that took place in Australia from the 1990s has resulted in a holistic approach to understanding and managing the Australian environment, including the introduction of modern ecological principles and provisions that are able to assess the ecological limitations of the soil environment and is based on sharing the responsibility for soil conservation between public and private interests.[35]

4 Commonwealth Level

Australia is a federation of six states and two territories. In general, the responsibility for land use decision-making and hence, historically, environmental protection, has lain with state governments. When the Commonwealth of Australia Constitution Act was passed in 1900, environmental protection was not an issue which occupied the minds of the legislators, and since then proposals to insert an "environmental" head of power into the Constitution via a referendum have not been pursued. There are

---

[33]The Australian Soil Resource Information System provides access to information and data products for a number of themes. Themes are developed to assist users to quickly view data of a topic of interest without having to negotiate the many data layers on the ASRIS maps page. Nutrient Management, Atlas of Australian Soils, Physiographic Regions of Australia, Acid Sulfate Soils and Crop Modelling are available.

[34]The 1989 law was repealed and the soil and water conservation provisions were replaced by the introduction of the Natural Resources Management Act 2004.

[35]E.g. the New South Wales Local Land Services Act 2013; Section 36 provides for the preparation of the state strategic plan which sets the vision, priorities and overarching strategy for local land services in the state, with a focus on appropriate economic, social and environmental outcomes.

many aspects of the distribution of power in the Australian federal system, in particular, the scope of Commonwealth legislative power, when the Commonwealth takes action in its federal capacity, which has a role and influence in the implementation of soil conservation activities.

4.1   Australian Constitution

The Constitution is the sole instrument which distributes legislative power within Australia. It was designed to foster the national interest and, historically, the creation and development of national government have been a response to the need to deal with national issues. On Federation in 1900, the Constitution provided that the general power to legislate continued to operate unless the Commonwealth was given exclusive power or the Commonwealth exercised its concurrent powers in which case Section 109 of the Constitution provides that the Commonwealth law prevails. In short, the Commonwealth was given specific powers, and the states were left with a single, undefined corpus of residential power. The distribution of power is determined by interpreting the words conferring specific powers upon the Common-wealth which are given their full and natural meaning.[36] Most Commonwealth powers are concurrent, which allows for continuity through the continued use of State power but enables the Commonwealth to enter and take over areas of power as issues of national concern. However, the Commonwealth's powers are limited, but those limits are found by interpreting the scope of the Commonwealth powers, there being no specific State powers by reference to which their limits can be determined. This is the most fundamental single premise of constitutional interpretation so far as the distribution of legislative power is concerned for the natural environment, and in this regard, determines the role of the Commonwealth and states in relation to soil conservation and natural resource management generally. Over time, the states have established very effective lines of communication with the Commonwealth government on soil conservation, this being the reason why the states have had specific and the bulk of legislation in this area.[37]

4.2   Purpose of Commonwealth Laws

With regard to policy, and the national interest, the Commonwealth has a clear responsibility to the management of soil as it is a natural resource of substantial national interest, although various Commonwealth government-level policies

---

[36]See *Amalgamated Society of Engineers v Adel SS Co Ltd* (1920) 28 CLR 129, *Uther v Federal Commissioner of Taxation* [1947] 74 CLR 508.

[37]Australia (1978a) and Hannam (2006).

implemented over the years have been to the detriment of land stability across the country (e.g. land settlement policies, financial incentives for land development).[38] The principle under the Constitution is that legislative powers are given to the Commonwealth only for certain purposes. Powers are typically conferred to legislate "with respect to", or about certain matters and not in achieving certain purposes (e.g. taxation-related policy was used in the past to break up larger holdings, promote closer settlement, to foster the settlement and development of rural land). When the Commonwealth enacts a law, it brings that law within the scope of one or more heads of power. For example, where the Commonwealth has the power to legislate with respect to trading corporations, it may directly control the activities of such corporations, including their land degrading activities. The important fact is that the Commonwealth has considerable power to deal with rural and land use matters, although they are not expressly mentioned in the Constitution.[39]

In Australia, whenever the role of the Commonwealth in the environment arises, the debate usually quickly turns to which level of government can better perform the tasks of devising and implementing the various aspects of environmental responsibility. There have been a number of situations where constitutional powers have been applied to "intervene" in a State environmental dispute, indicating that the Commonwealth has significant power to act in the interests of a national environmental problem as a whole. However, one of the main determinants of legislative involvement is the definition of "environment".[40] Australia has responded by introducing numerous specialist environmental laws at the Commonwealth and State level in response to the global concerns of biodiversity, desertification, threatened species protection, rangeland ecology, and more recently, climate change.[41]

4.3 Role of the Commonwealth

Because of the points mentioned above, the Commonwealth has not taken direct, formal legislative action for soil conservation, as it has always considered that the combined action of the states in this area has been "in the national interest"—with the emphasis of the state laws being placed on the nature, extent and seeking ways to deal with land degradation and soil conservation adequately. By choice, the

---

[38] Australia (1984, 1989); Bradsen (1988), pp. 143–149.

[39] Bradsen (1988), p. 133.

[40] Section 528 of the Commonwealth Environment Protection and Biodiversity Conservation Act 1999 defines "environment as including (a) ecosystems and their constituent parts, including people and communities; and (b) natural and physical resources; and (c) the qualities and characteristics of locations, places and areas; and (d) heritage values of places; and (e) the social, economic and cultural aspects of a thing mentioned in paragraph (a), (b), (c) or (d).

[41] Hannam and Boer (2004); Commonwealth Environment Protection and Biodiversity Conservation Act 1999, Carbon Credits (Carbon Farming Initiative) Act 2011, Climate Change Authority Act 2011.

Commonwealth has, in effect, taken a national coordinating and backstopping role, including

- A major role in the formulation of national natural resources management policy, including soil conservation policy.
- Providing a national coordinating administrative role (e.g. Australian Standing Committee on Soil Conservation—formed in the 1950s; National Soil Conservation Program—formed in 1983.
- Implementing the Natural Heritage Trust Program of 1997.
- Taking a major role in research (e.g. National Collaborative Study on Soil Conservation 1975-77[42]; under the Natural Heritage Trust 1997).
- Providing substantial financial assistance (e.g. through grants, loan money, National Soil Conservation Program).[43]
- Implementing the national Landcare program and Caring for Country Program.

4.4 Commonwealth Legislative Powers Relevant to Soil Conservation

While the Commonwealth has taken a prominent role in national policy, coordination and strategic aspects of soil conservation, over the years, various specific powers under the Constitution have been utilised in the interests of soil conservation and land degradation control. The taxation power has been used to allow taxation deductions for soil conservation since the 1940s.[44] This power could be used more positively to offer rebates or set differential rates of tax, including high rates of tax for activities which degrade the soil. The power to legislate with respect to bounties could be used to similar effect to that of tax deductions or rebates. Bounties can be paid on equipment or other inputs which are used to promote soil conservation.[45] Banks (which are a specific type of financial corporation; sections 51(13) insurance; and 51(14) banking, of the Constitution) can be required to take into consideration the provision of soil conservation in lending, or to give interest rate advantages to land users to help control the effects of soil degradation, drought and flooding.

Section 96 of the Constitution is a significant source of Commonwealth influence in the context of soil conservation activities and soil management. The Commonwealth can make grants to the states, giving it overwhelming dominance over the raising of public money whether by way of direct and indirect taxation or loans. Section 96 grants have a long history of use in the agricultural area, subsidising

---

[42]Collaborative Study (1978a) Report 3 Towards a National Approach to Land Resource Appraisal; Report 12, A Register of Current Australian Soil Conservation Research.

[43]E.g. Commonwealth Natural Resources Management (Financial Assistance) Act 1992.

[44]The Constitution Section 51(2).

[45]The Constitution Section 51(3).

wheat growers as early as the 1930s. Recognition was given in 1974–75 for grants for soil conservation purposes, conditional on the States agreeing to participate in the Soil Conservation Collaborative Study. The states were also funded for the National Soil Conservation Project, which was the major outcome of the Collaborative Study.[46]

5 Categorization of Australian Laws Relevant to Soil Conservation

An investigation of the Australian legal database determines that there are around 204 consolidated laws in the nine Australian jurisdictions that are relevant to soil conservation. Of these, 12 are in the Commonwealth jurisdiction and 192 in the states. The 204 laws are categorised into three classes, depending on their assessed role in soil conservation activity.[47] The three categories are

- Principal soil conservation laws.
- Main supporting legislation to the soil conservation objective.
- General support legislation to the soil conservation objective.

Table 1 is a breakdown of the laws according to the nine jurisdictions.[48]

By the early 1990s, in keeping with the processes of the Intergovernmental Agreement on the Environment, most States had commenced comprehensive environmental law and policy reform in the areas of water, vegetation, environmental assessment, land planning, and pollution management, with a major focus on integrated natural resources legislation. Although a number of States still have a specific "soil conservation" law, most of which are more than 30 years old,[49] the growing trend has been to either abolish the original laws and absorb soil conservation responsibilities within various forms of integrated natural resource laws or to create new, additional, comprehensive integrated natural resources law that make provision for soil conservation as a more general form of resource management activity. From the analysis, the main characteristic of the existing legal system for soil conservation in Australia is that each jurisdiction has a number of primary

---

[46]Collaborative Study (1978a) Report 1, A Basis for Soil Conservation Policy in Australia.

[47]The respective laws in the three categories contain legal rules (a rule externally compels a person, through force, threat or punishment, to do the things that the respective governments have deemed good or right in terms of soil conservation) and principles (to motivate people to do the things that seem good and right in terms of soil conservation. Further, various laws provide for the development of programs (a plan, action or schedule of activities, procedures, etc., to be followed that will achieve soil conservation).

[48]Original table was from Hannam (2006) but updated in 2018.

[49]Queensland Soil Conservation Act 1986; New South Wales Soil Conservation Act 1938; Western Australian Soil and Land Conservation Act 1945; Northern Territory Soil Conservation and Land Utilisation Act 1980.

Table 1 Breakdown of Australian laws relevant to soil conservation according to the nine jurisdictions

Jurisdiction	Principal SC law[a]	Main supporting legislation to SC objective[b]	General supportive legislation to SC objective[c]	Total
Commonwealth	None	7	5	12
Queensland	1 (specific SC law)	8	4	13
New South Wales	1 (specific SC law)	15	18	34
Victoria	2[d]	14	13	29
South Australia	2[e]	16	16	34
Western Australia	2[f] (1 specific SC law)	15	9	26
Tasmania	1[g]	13	9	23
Northern Territory	1 (specific SC law)	8	16	25
ACT[h]	None	6	2	8
Total	10	102	92	204

[a]Specific legislation for soil conservation
[b]Legislation that provides directly for specific soil conservation functions and activities
[c]Legislation that provides for functions which indirectly contribute to the soil conservation goal, objective or activity to be achieved
[d]Includes the Catchment and Land Protection Act 1994, also, there is provision for soil conservation in the Conservation, Forests and Lands Act 1987
[e]Soil conservation legislative provisions included in Natural Resources Management Act 2004
[f]Soil conservation legislative provisions included in Conservation and Land Management Act 1984
[g]Soil conservation legislative provisions included in Natural Resource Management Act 2002
[h]ACT—Australian Capital Territory

environmental laws that are now associated with natural resources management. In general, there is at least one principal environmental law that has a major coordination role, and this law generally sets the rules and standards that many of the individual pieces of legislation are obligated to follow in land, vegetation and general environmental management.[50]

Each individual jurisdiction has a number of primary environmental laws that establish the standards, rules, policies, auditing and compliance responsibilities for natural resources management (i.e. sustainable land management; ecologically sustainable development; integrated natural resource management; biodiversity

---

[50]The current system of integrated natural resource law is very human resource efficient for Australia, a large country with a small population base, as it enables groups of natural resource specialists to work closely together to identify and solve natural ecosystem management problems which are characterised by complex and multidisciplinary environmental issues. Concomitant with the evolution of integrated natural resources management law has been the reorganisation of government institutions—moving from the individual specialist institutions of the past, to fewer integrated multi-function environmental institutions with a wide range of scientific, ecological, economic, sociological and technical expertise.

management) and in general, there is one principal law that has a major coordination role and often sets the rules and standards that a number of the single pieces of legislation are obliged to follow.[51] Under these circumstances, the "single issue" legislation for soil, water, forestry and vegetation, play a supplementary role to the responsibilities of the primary environmental laws, where they set out basic natural resource management standards for these single resource areas which are then implemented through the functions of the primary environmental law mechanisms. There is good evidence that soil conservation law was used to control soil erosion to support the objectives for sustained agricultural production, as against being used for more complex objectives of land degradation control in a holistic environmental management context. There was also a long-term tradition that soil conservation law came under the administration of agriculture or primary industry government agencies rather than broader-based conservation organisations.[52]

By 2018, five of the eight Australian states still had an operational soil conservation law—Category 1 laws, but these laws, other than the Victoria Catchment and Land Protection Act 1994, are very much overshadowed by the supporting environmental laws for natural resources management as depicted in Category 2. Four of these five laws are special "soil conservation" laws and are structured along the lines of the original soil conservation laws introduced, i.e.

- Queensland Soil Conservation Act 1986.
- New South Wales Soil Conservation Act 1938.
- Western Australian Soil and Land Conservation Act 1945.
- Northern Territory Soil Conservation and Land Utilisation Act 1980.

The Victorian Catchment and Land Protection Act 1994 is an exception, in that it includes a formal procedure for managing catchments. The main purpose of this law is managing land and water resources and preventing land degradation.[53] In this chapter, a principal soil conservation law is defined as a law implemented to control soil erosion, and in some cases, land degradation processes (e.g. salinity, nutrient decline). While no new specific soil conservation law has been enacted in Australia since 1986 (the Queensland law was the most recent), a number of jurisdictions have introduced, in conjunction with an existing soil conservation law, a natural resources law that provides for the responsibilities of the original individual soil conservation law, including

---

[51]E.g. at the Commonwealth level it is the Environment Protection and Biodiversity Conservation Act 1999.

[52]One exception was the implementation of forestry, soil conservation and water conservation in New South Wales under the Conservation Authority Act 1949, where the Conservation Authority of New South Wales had powers, authorities, duties and functions to coordinate the activities of the three agencies.

Hannam (2000, 2003).

[53]Under Section 3 of the Victorian Catchment and Land Protection Act 1994 the meaning of "land" includes soil, water, vegetation and fauna on land.

- Western Australian introduced the Conservation and Land Management Act in 1984.[54]
- South Australia introduced the Natural Resources Management Act in 2004.[55]
- Tasmania introduced the Natural Resource Management Act in 2002.[56]

Examination of the individual laws in the three categories of legislation relevant to soil conservation indicates that there are numerous legal and legal-based mechanisms that enable soil conservation activities. There are many instruments that are applied for a specific soil conservation activity (e.g. construction of farm water supplies, soil conservation works program), but there are also many instruments that are applied for a particular environmental management function that has a major benefit for soil conservation (e.g. a reforestation program; reclamation of a mining area; protecting vegetation associations as habitat of an endangered species).

5.1 Category 1: Principal Soil Conservation Legislation

The investigation of the Australian legal database indicates that there are 10 individual laws across the nine jurisdictions that have an essential role in achieving the soil conservation objective. Legislation in this category includes the specific, traditional, soil conservation laws, and a few more recently introduced integrated natural resource laws in particular States where these laws have replaced the former specific soil conservation law.[57] Australia has a long history of legislation for the express purpose of soil conservation.[58] Each Australian state (except the small island State of Tasmania which has never had a specific soil conservation law of any form) has had some form of soil conservation law from the early 1930s to at least the 1980s. These laws were used with varying degrees of success, mainly dependent upon the prominence that agricultural production has played in the macro-economic development of Australia. By the early 1990s, in keeping with the national agenda of environmentalism (consolidated in the nation's Intergovernmental Agreement on the

---

[54]This law operates jointly with the Western Australia Soil and Land Conservation Act 1945.

[55]South Australia introduced a Sand Drift Act in 1923, followed by the Soil Conservation Act in 1939, the Soil Conservation and Land Care Act in 1989 (in conjunction with the Pastoral Land Management and Conservation Act of 1989), and finally, the Natural Resources Management Act 2004.

[56]Under Section 3 of this law, "natural resource management" means management of any activity that uses, develops or conserves (a) air, water, land, plants, animals and micro-organisms; and (b) the systems they form; under Schedule 1 of this law, the objectives of the resource management and planning system is to provide for the fair, orderly and sustainable use and development of air, land and water.

[57]Hannam (2006), the objectives of each individual law in Category 3 are summarised in Appendix 2.

[58]Bradsen (1988) and Hannam (2000, 2006).

Environment),[59] most states had commenced comprehensive environmental law and policy reform in the areas of water, vegetation, environmental assessment, environment protection, land planning, and pollution management and with a greater focus on integrated natural resources legislation. The traditional "soil conservation" law was not a prominent part of this Australia-wide reform, and the eight states reacted as follows:

- Four states—Queensland, New South Wales, Western Australia and Northern Territory, have retained specific soil conservation laws, but have introduced additional, integrated natural resource management laws that carry many soil conservation functions.
- Two states—Victoria and South Australia abolished their specific soil conservation laws and introduced comprehensive integrated natural resources law, making provision for soil conservation issues as a general component of broad natural resource management activity.
- The Australian Capital Territory previously did not have a specific soil conservation law, but a Soil Conservation Ordinance, introduced in 1960. Now, soil conservation responsibilities are part of a comprehensive environmental regime which includes the Environment Protection Act 1997, the Land (Planning and Environment) Act 1991 and the Planning and Land Act 2001.
- The State of Tasmania has never had a soil conservation instrument. The Natural Resource Management Act 2002 is responsible for land and water management.

Despite the growth of integrated natural resource laws, the specific soil conservation laws of Category 1 have many elements that remain important to achieve the general objective of soil conservation, and therefore, the broader environmental agenda. However, it must be recognised that, alone, these specific laws do not have the full range of environmental law elements required to fully achieve the environmental objectives of soil conservation—as sought under the main national and State environmental and conservation strategies.

Some of the key elements of the four current specific soil conservation laws include[60]

- Provision to set up some form of "soil" institution.
- The appointment of a departmental head.
- The establishment of advisory committees.
- Soil conservation planning provisions (soil survey, farm water supplies, catchment planning and farm planning).
- The establishment of soil conservation schemes of works (practical soil and farm water conservation projects).
- The development of agreements between the State and individuals to implement soil conservation works (usually providing some form of financial assistance).

---

[59]Australia (1992a).

[60]Bradsen (1988); Hannam and Boer (2002) Section 4, p. 33.

- The acquisition of land for special soil conservation restoration projects.
- The enforcement and compliance procedures to rectify practical soil conservation problems.

5.1.1 Main Elements of Other Principal Laws with Primary Responsibility for Soil Conservation

The main aim of other principal laws with primary responsibility for soil conservation, including

- Victoria Conservation, Forests and Lands Act 1987.
- South Australia Natural Resources Management Act 2004 and the Pastoral Land Management and Conservation Act 1989.
- Western Australia Conservation and Land Management Act 1984.
- Tasmania Natural Resource Management Act 2002.
    is to achieve sustainable land management[61] by establishing an integrated legislative scheme to promote the use and management of natural resources in a manner that:
- Recognizes and protects the intrinsic values of soil resources.
- Protects biological diversity and encourages the restoration or rehabilitation of soil ecological systems and processes that have been lost or degraded.
- Provides for the protection and management of catchments and the sustainable use of land and seeks to restore land resources that have been degraded.
- Supports sustainable primary and other economic production systems with particular reference to the value of agriculture to the economy.
- Provides for the prevention or control of impacts caused by species of animals and plants that may have an adverse effect on the environment, primary production or the community.
- Promotes educational initiatives and provides support mechanisms to increase the capacity of people to be involved in the management of soil resources.

In these laws, the term "sustainable land management" comprises the use, conservation, development and enhancement of natural resources in a way, and at a rate, that will enable people and communities to provide for their economic, social and physical well-being while[62]:

- Sustaining the potential of natural resources to meet the reasonably foreseeable needs of future generations.
- Safeguarding the life-supporting capacities of natural resources.
- Avoiding, remedying or mitigating any adverse effects of activities on natural resources.

---

[61]This term is widely used in the principal laws.

[62]E.g. Section 7 South Australia Natural Resources Management Act 2004; Section 7 Tasmania Natural Resource Management Act 2002.

5.2 Category 2 Main Supporting Legislation to Soil Conservation Objectives

The investigation of the Australian legal database indicates that there are at least 102 individual laws across the nine jurisdictions that have an essential role in achieving the soil conservation objective. The legislation in this category includes the natural resource laws that establish primary standards, rules, planning, evaluation, policy-making, and auditing and compliance mechanisms for all aspects of environmental management in Australia. Under these arrangements, soil conservation legislation has a supplementary role in providing the basic standards for soil conservation activities and functions as a key part of this overall environmental management regime in Australia.[63]

5.2.1 Category 2 Laws: Commonwealth Level

At the Commonwealth level, the main supporting laws relevant to soil conservation include:

- National Environment Protection Council Act 1994.
- National Environment Protection Measures (Implementation) Act 1998.
- Environment Protection and Biodiversity Conservation Act 1999.
- Natural Heritage Trust of Australia Act 1997.
- Natural Resources Management (Financial Assistance) Act 1992.
- Water Act 2007.
- Climate Change Authority Act 2011.[64]

5.2.2 National Protection of the Environment and Biodiversity

The Commonwealth Environment Protection and Biodiversity Conservation Act 1999 and the Natural Heritage Trust of Australia Act 1997 are important Commonwealth laws relevant to national soil conservation objectives. In particular, the

---

[63]Hannam (2006), the objectives of each individual law in Category 2 are summarized in Appendix 2.

[64]Under Section 11(c), the Climate Change Authority can conduct research about matters relating to climate change. Further, under the Carbon Credits (Carbon Farming Initiative) Act (2011), and Section 3 (2). The first object of this Act is to remove greenhouse gases from the atmosphere, and avoid emissions of greenhouse gases, in order to meet Australia's obligations under any or all of the following: (a) the Climate Change Convention; (b) the Kyoto Protocol; (c) an international agreement (if any) that is the successor (whether immediate or otherwise) to the Kyoto Protocol. (3) The second object of this Act is to create incentives for people to carry on certain offsets projects. (4) The third object of this Act is to increase carbon abatement in a manner that: (a) is consistent with the protection of Australia's natural environment; and (b) improves resilience to the effects of climate change.

Environment Protection and Biodiversity Conservation Act 1999 provides for the protection of the environment, especially those aspects of the environment that are matters of national environmental significance. It promotes ecologically sustainable development through the conservation and ecologically sustainable use of natural resources and promotes the conservation of biodiversity. It is implemented on a co-operative basis with the states to manage the environment with the involvement of governments, the community, land holders and Indigenous peoples. Importantly, it assists in the co-operative implementation of Australia's international environmental responsibilities and recognizes the role of Indigenous people in the conservation and ecologically sustainable use of Australia's biodiversity.[65] It promotes the use of Indigenous peoples' knowledge of biodiversity and land management with the involvement of, and in co-operation with, the owners of the knowledge. The Environment Protection and Biodiversity Conservation Act 1999 is based on the principles of ecologically sustainable development and its decision-making processes effectively integrate both long-term and short-term economic, environmental, social and equitable considerations. Importantly, this law is structured on three key principles of international environmental law, being[66]:

- The precautionary principle—if there are threats of serious or irreversible environmental damage, lack of full scientific certainty should not be used as a reason for postponing measures to prevent environmental degradation.
- The principle of intergenerational equity—that the present generation should ensure that the health, diversity and productivity of the environment are maintained or enhanced for the benefit of future generations.
- The principle of biological diversity—the conservation of biological diversity and ecological integrity should be a fundamental consideration in decision-making.

The Natural Heritage Trust of Australia Act 1997 established the Natural Heritage Trust of Australia Account, which is the main repository of the AUD1.35 billion from the partial sale of Australia's public communication organisation. The main

---

[65]The Commonwealth Environment Protection and Biodiversity Conservation Act 1999 introduce into Australian law many aspects of key global treaties, including the 1992 Biological Diversity Convention; 1992 United Nations Framework Convention for Climate Change, 1995 United Nations Convention to Combat Desertification; Note, "biological diversity" is defined in various state laws e.g. in Section 10 of the Queensland Nature Conservation Act 1992 it is defined as (1) 'the natural diversity of native wildlife, together with the environmental conditions necessary for their survival, and includes (a) regional diversity, that is, the diversity of the landscape components of a region, and the functional relationships that affect environmental conditions within ecosystems; and (b) ecosystem diversity, that is, the diversity of the different types of communities formed by living organisms and the relations between them; and (c) species diversity, that is, the diversity of species; and (d) genetic diversity, that is, the diversity of genes within each species'. Section 10(2) specifies that 'in subsection (1) landscape components includes landforms, soils, water, climate, wildlife and land uses'.

[66]Hannam and Boer (2004) Section II, International principles for drafting national soil legislation, p. 21.

objective of the account is to provide the funds, and therefore the major strategy, to conserve, repair and replenish Australia's natural capital infrastructure. This law finances the major national expenditure on the Australian environment, particularly for sustainable agriculture and natural resources management. The Natural Heritage Trust Advisory Committee established under this law is responsible for managing this law and the account which funds various major national natural resource programs which are crucial for sustainable management of soil.[67]

5.3   Category 2: State-Level Laws

At the state level, there are around 95 individual laws that establish primary standards, rules, and procedures for natural resource evaluation, resource policy-making, auditing and compliance—that have a significant effect on soil conservation activity. Of these 95 laws, around 60 are considered the most relevant to the objective of soil conservation, and include, for example

Queensland

- Nature Conservation Act 1992.
- Planning Act 2016.
- Water Act 2000.

New South Wales

- Biodiversity Conservation Act 2016.[68]
- Crown Land Management Act 2016.[69]
- Environmental Planning and Assessment Act 1979.
- Farm Water Supplies Act 1946.
- Local Land Services Act 2013.
- Murray–Darling Basin Act 1992.
- National Environment Protection Council (New South Wales) Act 1995.
- Water Act 2014.
- Western Lands Act 1901.

---

[67]Environmental protection is defined under Section 15, sustainable agriculture is defined under Section 16, and natural resources management is defined under Section 17 of the Natural Heritage Trust of Australia Act 1997.

[68]Section 1.3 the purpose of this Act is to maintain a healthy, productive and resilient environment for the greatest well-being of the community, now and into the future, consistent with the principles of ecologically sustainable development (described in Section 6(2) of the Protection of the Environment Administration Act 1991).

[69]Under Section 1.3(c) the object is to require environmental, social, cultural heritage and economic considerations to be taken into account in decision-making about Crown land.

Victoria

- Environment Protection Act 1970.
- Land Act 1958.
- Murray-Darling Basin Act 1993.
- National Environment Protection Council (Victoria) Act 1995.
- Planning and Environment Act 1987.
- Victorian Environmental Assessment Council Act 2001.
- Water Act 1989.
- Water Industry Act 1994.

South Australia

- Environment Protection Act 1993.
- Murray-Darling Basin Act 2008.
- National Environment Protection Council (South Australia) Act 1995.
- Native Vegetation Act 1991.
- River Murray Act 2003.
- Water Resources Act 1997.

Western Australia

- Agriculture and Related Resources Protection Act 1976.
- Biosecurity and Agriculture Management Act 2007.
- Biodiversity Conservation Act 2016.
- Environmental Protection Act 1986.
- Planning and Development Act 2005.
- Water Resources Legislation Amendment Act 2007.

Tasmania

- Environmental Management and Pollution Control Act 1994.
- Farm Water Development Act 1985.
- Land Use Planning and Approvals Act 1993.
- National Environment Protection Council (Tasmania) Act 1995.
- Nature Conservation Act 2002.
- Water Management Act 1999.

Northern Territory

- Environmental Assessment Act 2013.
- Environmental Offences and Penalties Act 1996.
- Pastoral Land Act 2016.
- Planning Act 2017.
- Water Act 2016.

Australian Capital Territory

- Planning and Development Act 2007.
- Native Title Act 1994.
- Nature Conservation Act 2014.
- Water Resources Act 2007.

5.3.1 Administration

The Category 2 state laws are administered by state organizations or authorities which in some jurisdictions also administer the specialist soil conservation law. In this regard, there is a wide variety of administrative bodies that have a role in the implementation of the Category 2 laws, including state councils, regional boards, catchment management authorities, commissions, corporations, specialist authorities, advisory boards, advisory committees, environmental trusts, and technical working groups.[70] The powers, duties and authorities of these administrative bodies carry a wide range of functions relevant for soil conservation including policy-making, approving pastoral and water licenses, granting permission to undertake resource management activities within prescribed limits, decision-making and issuing directions for soil utilisation.

5.3.2 Soil Conservation Benefits

This group of legislation has a role in developing natural resource standards, limits and conditions of use, which include many soil conservation measures. New South Wales has a sophisticated approach by virtue of the Natural Resources Commission Act of 2003. This law establishes an independent body with broad investigating and reporting functions. Its purpose is to establish a sound ecological basis for the management of natural resources in the social, economic and environmental interests of the State, with the adoption of State-wide standards and targets for natural resource management issues.[71]

---

[70]E.g. the functions of Local Services Boards formed under Section 26 of the New South Wales Local Land Services Act 2013, is 'to determine the general policies and strategic direction of Local Land Services' (Section 4). Under Section 4(1) 'local land services', means programs and advisory services associated with agricultural production, biosecurity, natural resource management and emergency management, including programs and advisory services associated with the following: (a) agricultural production, (b) biosecurity, including animal pest and disease and plant pest and disease prevention, management, control and eradication, (c) preparedness, response and recovery for animal pest and disease and plant pest and disease emergencies and other emergencies impacting on primary production or animal health and safety, (d) animal welfare, (e) chemical residue prevention, management and control, (f) natural resource management and planning, (g) travelling stock reserves and stock watering places, (h) control and movement of stock, (i) related services and programs.

[71]Section 3; under Section 5, "natural resource management" extends to the following matters relating to the management of natural resources: (a) water, (b) native vegetation, (c) salinity, (d) soil, (e) biodiversity.

5.3.3 Courts and Tribunals

There are a range of laws that provide for land use issues, including soil conservation issues, to be heard. The main laws include:

- South Australia Environment, Resources and Development Court Act 1993.
- Tasmania Resource Management and Planning Appeal Tribunal Act 1993.
- Australian Capital Territory Commissioner for Sustainability and the Environment Act 1993.
- Western Australia Agriculture and Related Resources Protection Act 1976.
- New South Wales Land and Environment Court Act 1979.
- Queensland Land Court Act 2000.

Under the Australian Capital Territory Commissioner for Sustainability and the Environment Act 1993, the Commissioner can investigate complaints regarding the management of the environment by the Territory or a Territory authority.[72] The Commissioner can conduct investigations into actions of an agency whose actions would have a substantial impact on the environment of the Territory. There is also a power to prepare a state of the environment report which can include an assessment of the condition of the environment. The assessment will consider the components of the earth, including soil, the atmosphere and water, any organic or inorganic matter and any living organism, ecosystems and their constituent parts— including people and communities.[73] Other relevant aspects that may be considered include the qualities and characteristics of places and areas that contribute to their biological diversity and ecological integrity, scientific value and amenity. A state of the environment report may also include an evaluation of the effectiveness of environmental management, including an assessment about the degree of compliance with national environmental protection measures made by the National Environment Protection Council under the Commonwealth National Environment Protection Council Act 1994 and National Environment Protection Measures (Implementation) Act 1998.

5.4   Category 3 General Supportive Legislation to Soil Conservation Objective

The investigation of the Australian legal database indicates that there are 92 laws across the nine jurisdictions that have a general supportive role in achieving the soil conservation objective. It includes legislation with functions that indirectly contribute to, or enable the achievement of a soil conservation goal, objective or activity. Unlike Category 2 legislation which has an overall role in environmental

---

[72]Section 13.
[73]Section 15.

management, the legislation in Category 3 is generally indirectly concerned with the management of land resources in some supportive way.[74]

At the Commonwealth level, relevant laws include

- United Nations Food and Agriculture Organization Act 1944.
- Rural Adjustment Act 1992.
- Native Title Act 1993.
- Native Title Amendment Act 1998.

The Rural Adjustment Act 1992 provides benefits on both a national basis and a regional basis by fostering the development of a profitable farm sector that is able to operate competitively in a deregulated financial and market environment. It aims to improve the competitiveness of the farm sector by assisting in the application of sustainable land management measures. Such measures include soil conservation measures that lead to improved farm production. The major intention of this law is to promote a better financial, technical and management performance from the farm sector, and support, either directly or indirectly, farmers who have prospects of sustainable long-term profitability with a view to improving the productivity of their farm units.[75] Support is provided in a way that ensures that farmers become financially independent of that support within a reasonable period. This is in the form of grants and subsidies for interest and costs payable on loans whether the loans are provided by a state or by another person. Grants are also available for farm training, planning, appraisal, support services and rural adjustment research. Support is also given to farmers who do not have prospects of sustainable long-term profitability to leave the farm sector and enables grants of money to be made to persons other than farmers for purposes relating to rural adjustment.[76]

The Native Title Act 1993 is relevant to soil conservation because it provides for the recognition and protection of Indigenous land use, which occupies a very large area of mainly pastoral land, in Australia. This law helps establish ways in which future dealings affecting native title may proceed, and sets standards for those dealings. It also establishes a mechanism for determining claims to native title and permits the validation of past acts invalidated because of the existence of native title.[77]

5.4.1   Category 3: State Level Laws

At the State level, there are 87 laws that establish land use standards, rules, planning, evaluation, policy-making, and auditing and compliance mechanisms that enable soil conservation measures to be implemented. Category 3 laws cover such areas of

---

[74]Hannam (2006), the objectives of each individual law in Category 3 are summarized in Appendix 2.
[75]Section 3(1).
[76]Section 3(2).
[77]Section 3.

interest as—native title, bushfire control, mining, agriculture land tenancy, forestry, conservation trusts, education, nature conservation, primary industry funding and carbon trading. Of the 87 laws, around 70 are considered the most relevant to soil conservation and include, for example

Queensland

- Mineral Resources Act 1989.
- Dispute Resolution Centres Act 1990.
- Environmental Protection Act 1994.
- Rural and Regional Adjustment Act 1994.

New South Wales

- National Parks and Wildlife Act 1974.
- Land and Environment Court Act 1979.
- Wilderness Act 1987.
- Agricultural Tenancies Act 1990.
- Mining Act 1992.
- Native Title (New South Wales) Act 1994.
- Forestry Restructuring and Nature Conservation Act 1995.
- Rural Fires Act 1997.
- Environmental Trust Act 1998.
- National Park Estate (Reservations) Act 2003.

Victoria

- Forests Act 1958.
- Victorian Conservation Trust Act 1972.
- National Parks Act 1975.
- Victorian College of Agriculture and Horticulture Act 1982.
- Flora and Fauna Guarantee Act 1988.
- Mineral Resources (Sustainable Development) Act 1990.
- Heritage Rivers Act 1992.

South Australia

- Law of Property Act 1936.
- Rural Advances Guarantee Act 1963.
- Primary Producers Emergency Assistance Act 1967.
- Land Acquisition Act 1969.
- Mining Act 1971.
- National Parks and Wildlife Act 1972.
- Groundwater (Border Agreement) Act 1985.
- Rural Industry Adjustment and Development Act 1985.
- Development Act 1993.
- Economic Development Act 1993.
- Primary Industry Funding Schemes Act 1998.
- South Australian Forestry Corporation Act 2000.

- Lake Eyre Basin (Intergovernmental Agreement) Act 2001.
- Irrigation Act 2009.

  Western Australia

- Wildlife Conservation Act 1950.
- Bush Fires Act 1954.
- Mining Act 1978.
- Bunbury Tree-farm Project Agreement Act 1995.
- National Environment Protection Council (Western Australia) Act 1996.
- Native Title (State Provisions) Act 1999.
- Forest Products Act 2000.
- Carbon Rights Act 2003.

  Tasmania

- Fire Service Act 1979.
- Forest Practices Act 1985.
- Forestry Rights Registration Act 1990.
- Private Forests Act 1994.
- Mineral Resources Development Act 1995.
- Primary Industries Activities Protection Act 1995.
- Regional Forest Agreement (Land Classification) Act 1998.
- Forest Practices (Private Timber Reserves Validation) Act 1999.
- Forestry Management Act 2013.

  Northern Territory

- Land Title Act 1980.
- Lands Acquisition (Pastoral Leases) Act 1982.
- Mining Management Act 2001.
- Weeds Management Act 2001.
- Parks and Reserves (Framework for the Future) Act 2003.
- Parks and Wildlife Commission Act 2013.
- Territory Parks and Wildlife Conservation Act 2013.
- Waste Management and Pollution Control Act 2016.

  Australian Capital Territory

- Enforcement of Public Interests Act 1973.
- Commissioner for Sustainability and the Environment Act 1993.
- Climate Change and Greenhouse Gas Reduction Act 2010.

5.4.2 Native Title

This area of law provides for Indigenous people to obtain rights to use vast areas of mainly arid and semi-arid lands where they can undertake pastoral and agricultural pursuits (e.g. New South Wales Native Title Act 1994, Western Australia Native

Title Act 1999). Most native title land being in arid and semi-arid environments means that it is ecologically fragile. This land requires the application of long-term soil conservation measures to maintain its productivity and ecological stability, including fire control, positioning of stock watering facilities, pasture improvement and native conservation management.

5.4.3 Bushfire Control

Wildfire in Australia is a frequent event and a significant causal agent of soil erosion, loss of soil fertility, loss of important vegetation species and associations, and ultimately land degradation. All states have legislative provisions relating to the use and management of fire. e.g. Western Australia Bush Fires Act 1954 and the New South Wales Rural Fires Act 1997.

The role of these laws is to provide for the prevention, mitigation and suppression of bushfires in local government areas and other parts of the state constituted as rural fire districts. They also provide for coordination of bushfire control activities and bushfire prevention measures. The more recent bushfire control laws provide for the protection of the environment by requiring that certain bushfire control activities, e.g. hazard reduction burning, are carried with regard to the principles of ecologically sustainable development, and in this regard, they benefit soil conservation objectives.

5.4.4 Mining

Various forms of mining activity lead to degradation of soils. All states have laws that control mining and extractive industries e.g.

- South Australia Mining Act 1971.
- Western Australia Mining Act 1978.
- Queensland Mineral Resources Act 1989.
- Victoria Mineral Resources (Sustainable Development) Act 1990.
- New South Wales Mining Act 1992.
- Tasmania Mineral Resources Development Act 1995.
- Northern Territory Mining Management Act 2001.

This area of law facilitates prospecting and exploring for mining of minerals while enhancing the knowledge of mineral resources and minimising land use conflict. It encourages environmental responsibility in mining activities and provides a framework to expedite and regulate mining with a land management responsibility. Under Australian law, most mining activities require approval under environmental protection laws, including the requirement to prepare an environmental impact assessment.[78] Most of the major environmental constraints to mining, and subsequent conditions for compliance, stem from the environmental assessment procedure, rather than procedures under the mining laws themselves.[79] However, in the Northern Territory, the Mining Management Act 2001 presides over the development of mineral resources in accordance with best practice safety, health and environmental standards. It authorises and monitors mining activities and requires the careful management of mining sites. This law sets out procedures for consultation and cooperation for environmental protection management systems, undertaking audits, inspections, investigations, monitoring and reporting to ensure compliance with agreed environmental standards and specifies the obligations in mining in respect of the environment. This law specifies the type of assistance available to the mining industry to introduce programs of continuous improvement and achieve best practice in environmental management.[80]

5.4.5 Agriculture Tenancy and Land Acquisition

There is provision for agricultural tenancy in various state lands law and related laws. The New South Wales Agricultural Tenancies Act 1990 specifically encourages agricultural landowners and their tenants and share farmers to employ the principles of ecologically sustainable development in their farming practices[81] and to maintain sustainable agricultural production and prevent degradation of the environment. This law provides for farmers to use written agreements under agricultural tenancy arrangements to set out their rights to farming and the terms of sustainable land use. It also provides a mechanism for resolution of disputes by the parties to agricultural tenancies themselves, through mediation, and an arbitration mechanism for settling disputes between parties to agricultural tenancies that is outside the court system.[82]

5.4.6 Nature Conservation

This area of law includes a wide range of nature conservation legislation including threatened species law, national parks and wildlife law, wilderness protection area law, and laws that establish and protect nature parks and reserves. The significant

---

[78]By definition, mining is an activity that requires approval under State Environmental Planning and Assessment laws and Environment Protection laws. A major consideration in assessment is the potential for protection of significant ecological values and successful land rehabilitation.

[79]E.g. see Section 6 Western Australia Mining Act 1978.

[80]Section 3(a)-(f); Part 3 Environmental obligations.

[81]Objects of the Act, Section 3; the principles of ecologically sustainable development as described by Section 6(2) of the Protection of the Environment Administration Act 1991.

[82]Part 4 of the Act, Dispute Resolution and Remedies.

aspect of this area of law to soil conservation is that in most cases the laws protect areas of land that are ecologically sensitive and would otherwise be highly susceptible to degradation if subjected to other forms of land use, particularly forestry, agriculture and grazing. Importantly, this area of legislation provides for the dedication of land for conservation protection purposes, but they also have comprehensive land assessment and land management requirements. In this regard, specific soil conservation measures are applied to protect soil resources and natural ecosystems.

The majority of these laws apply to public lands (including Crown lands). In some states, e.g. Victoria Flora and Fauna Guarantee Act 1988, Western Australia Wildlife Conservation Act 1950—this area of law will apply to private land management. In this regard, they are very relevant to soil conservation objectives as, often, land that is ecologically sensitive and with high biological diversity, quickly degrades if disturbed for agricultural or other intensive land use purposes. Thus, the protection of this land by law for conservation purposes prevents soil degradation. Soil conservation measures will be required on these lands as a part of basic land management, e.g. road maintenance, fire control, and revegetation.

6 Conclusions

There is no national law for soil conservation in Australia. The management of soil resources is primarily a state matter with most of the eight State jurisdictions having had some form of soil conservation law at some stage in the past. The national government takes a coordinating role through the development of national resource policies and strategies and controlling substantial financial resources which the states bid for under national funding guidelines. Five of the eight state jurisdictions have existing soil conservation laws, but these are greater than 30 years old and have not been amended for some time.

Using the standard Australian definition of soil conservation, it has been found that there are around 204 laws in Australia that are relevant to the implementation of soil conservation, as specified in the definition. The 204 laws are categorized into three categories according to their role in soil conservation, including; the specialized soil conservation laws, main supporting laws, and generally supporting laws. Five of the eight states have soil conservation laws, where four of the states still have a 'traditional' soil conservation law, and the state of Victoria has the Catchment and Land Protection Law 1994 which is a specialist soil and water conservation law. Although outdated by comparison to the modern environmental laws of Australia, the existing 'traditional' soil conservation laws have mechanisms to implement basic, practical soil conservation measures (e.g. land planning, land management, enforcement, advisory committees, declare areas of erosion hazard, research).

When the main era of modern environmental law reform took place in Australia in the 1990s, a number of states abolished their specific soil conservation laws and integrated soil conservation activities and responsibilities within comprehensive integrated natural resource laws. From this time, the remaining specialist soil conservation laws performed a subordinate role to principal integrated environmental laws. For these laws to remain effective, they require substantial reform and linking within Australia's modern integrated environmental law system. At this point, their role is unclear within the integrated system, a position that needs to be rectified. The structure of modern environmental laws has been greatly influenced by the introduction of international conventions for the environment and national strategies for ecologically sustainable development and integrated ecosystem management. The integrated environmental laws contain procedures for major land use decisions, land-use planning, allocation of land, integrated environmental policy, compliance and enforcement, dispute resolution, establishment of specialist environmental courts, financing environmental and conservation works e.g. Commonwealth Environment Protection and Biodiversity Conservation Act 1999, Western Australia Conservation and Land Management Act 1984, Tasmania Natural Resource Management Act 2002, New South Wales Natural Resources Commission Act 2003 and the South Australia Natural Resources Management Act 2004. The integrated natural resources legislation is implemented by multidisciplinary agencies or institutions which have responsibilities for water, forests, soils, and land management. One of the main characteristics of the integrated approach is the comprehensive role for the community in all aspects of land use decision-making and land management.

References

Australia (1978a) A Basis for Soil Conservation Policy in Australia, Commonwealth and State Government Collaborative Soil Conservation Study 1975–1977, Report 1. Department of Environment, Housing and Community Development

Australia (1978b) A Basis for Soil Conservation Policy in Australia, Commonwealth and State Government Collaborative Soil Conservation Study 1975–1977, Report 14, Legislative, Administrative and Financial Arrangement Affecting Soil Conservation. Department of Environment, Housing and Community Development

Australia (1983) National Conservation Strategy for Australia, Living Resource Conservation for Sustainable Development. Australian Government Publishing Service, Canberra

Australia (1984) Land use policy in Australia, Senate Standing Committee on Science, Technology and the Environment. Australian Government Publishing Service, Canberra

Australia (1989) The effectiveness of land degradation policies and programs, report of the house of representatives standing committee on environment, recreation and the arts. Australian Government Publishing Service, Canberra

Australia (1992a) Intergovernmental Agreement on the Environment

Australia (1992b) National strategy for ecologically sustainable development. Australian Government Publishing Service, Canberra

Australian Soil Conservation Council (1989) National Soil Conservation Strategy. Department of Primary Industries and Energy, Canberra

Bates G (2016) Environmental law in Australia. Butterworth's

Bradsen JR (1988) Soil conservation legislation in Australia, report for the National Soil Conservation Program. University of Adelaide

Breckwoldt R (1988) The dirt doctors, a jubilee history of the soil conservation service of NSW. Soil Conservation Service of NSW

Downes RG (1970) Soil conservation law in Australia. Soil Conservation Authority, Victoria

Graham OP (1992) Survey of land degradation in New South Wales, Australia. J Environ Manage 2:205

Hannam ID (2000) Soil conservation policies in Australia: successes, failures, and requirements for ecologically sustainable policy. In: Napier TL, Napier SM, Tvrdon J (eds) Soil and water conservation policies and programs: successes and failures. Soil and Water Conservation Society, CRC Press, New York

Hannam ID (2003) Soil conservation in Australia. J Soil Water Conserv 58(6):112A–115A

Hannam (2006) Soil and Water Conservation Law in Australia, Technical Assistance to the People's Republic of China for the Implementation of the National Strategy for Soil and Water Conservation, TA 4404. Asian Development Bank

Hannam ID, Boer BW (2002) Legal and institutional frameworks for sustainable soils, a preliminary report. IUCN, Gland, Switzerland and Cambridge, UK, 88 p

Hannam ID, Boer BW (2004) Drafting legislation for sustainable soils: a guide. IUCN, Gland, Switzerland and Cambridge, UK

Houghton PD, Charman PEV (1986) Glossary of terms used in soil conservation. Soil Conservation Service of New South Wales

Martin P (2017) Soils governance, an Australian perspective. In: Ginzky H et al (eds) International Yearbook of Soil Law and Policy 2016. Springer International Publishing

Annex: Outcome Document

International Workshop "Legal Instruments for the Effective Protection and Sustainable Management of Soils"

Organized in cooperation between the German Environment Agency (UBA), the Konrad-Adenauer Foundation—Climate Policy and Energy Security Program for Sub-Saharan Africa—Makerere University and Kampala International University
　Kampala, Uganda 26–27 September 2017
The Workshop on "Legal instruments for the effective protection and sustainable management of soils" was held at the Hotel Africana, Kampala, from 26th to 27th of September, 2017. The Workshop was attended by about 50 experts from African countries and abroad. Representatives from UNCCD, FAO, IUCN, UNEP, the African Soil Partnership and GIZ attended the Workshop.
　Keynote addresses were provided by;

- Ms. Petra Kochendörfer, Chargé d'Affaires a.i., German Embassy in Kampala;
- Prof. Dr. Oliver C Ruppel, Director of Konrad-Adenauer Stiftung (Climate Policy and Energy Security Program for Sub-Saharan Africa);
- Vincent Frerrio Bamulangaki Ssempijja, Hon. Minister of Agriculture, Animal Industry and Fisheries, Government of Uganda.

The Workshop sought to address several thematic issues of key relevance to the role of international, regional and domestic legal instruments in promoting the effective protection and sustainable management of soils. Below is a brief summary of the key content of these discussions.

Sustainable Management of Soils—General Aspects

1. The protection and sustainable management of soils is a precondition for sustainable development and more so for the survival of humankind. Without sufficient areas of fertile soils, there is no food security and no chance to mitigate climate

© Springer Nature Switzerland AG 2020
H. Yahyah et al. (eds.), *Legal Instruments for Sustainable Soil Management in Africa*, International Yearbook of Soil Law and Policy,
https://doi.org/10.1007/978-3-030-36004-7

change. Degraded soils result in hunger, famine, migration and, under certain circumstances, even in wars.

2. Land preservation and thus the sustainable management of soils is required to be able to achieve the majority of the Sustainable Development Goals of the UN 2030 Sustainable Development Agenda.

3. UNCCD has unanimously agreed on the following definition of "Land Degradation Neutrality" (LDN) which is crucial to determine the right actions to achieve sustainable management of soils:

LDN is "a state whereby the amount and quality of land resources necessary to support ecosystem functions and services and enhance food security remain stable or increase within specified temporal and spatial scales and ecosystems."

LDN is to be regarded as a key concept in this context.

4. From a *science* point of view, sustainable management of soils is a global concern. Statements and decisions by the international state community, e.g. the Rio+20 outcome document or during COP

12 UNCCD, also show that the sustainable management of soils Is politically regarded as common challenge of humankind. However, as yet this is not expressed in a corresponding *legal* principle such as common concern of humankind. Such a legal principle would clearly underline the need of more responsible and consolidated actions on all levels for a sustainable management of soils.

5. Although the effects of soil degradation are global, an appropriate management of soils must be implemented locally. The challenges are multifactorial including ecological, social, cultural, economic, political and legal aspects. Actions must therefore primarily be of local nature.

6. The initiative "Economics of Land Degradation" has shown that preventive measures are far more cost-effective than actions of restoration or rehabilitation.

Factual Challenges in Africa on Soils

7. African states face specific factual challenges concerning their soils and lands.

8. Although Africa the continent with the least land degradation, the pressure on soils is currently enormous and continuously increasing due to a range of factors including poverty, over-exploitation, population growth and climate change.

9. As the agricultural productivity is still low in Africa, food security has not yet been achieved.

10. Drivers of unsustainable soil management include overstocking, overgrazing, water erosion, landslides, and over-application of agro-chemicals. The poor population often depends on land and other natural resources for immediate needs which is an additional driver for land degradation.

11. Soils in Africa are inherently vulnerable due to low level of resilience.

12. The following aspects impose serious impediments to sustainable soil management in Africa:

- Data on soils are often not available or inadequate.
- Soil research results are often unavailable to farmers and are not sufficiently implemented in daily agricultural practices.

- The level of investment is often low.
- Sustainable soil management is not yet commonly observed as a crucial instrument to achieve sustainable development. It has not been imposed as a legal requirement in many aspects of African law.

Soil Protection Regulatory Concepts and Challenges in Africa

13. Based on the analysis of Ugandan and Namibian environmental law and a comparison of these laws with German regulatory approaches the following observations were made with regard to soil protection regulation in Africa.
14. There are usually several law and regulatory approaches being at least indirect relevant for soil protection. However, regulations are fragmented and/or do not follow a coherent and consistent policy.

- So far, sustainable soil management is not a legally established objective.
- There is no clear obligation to restore or rehabilitate degraded land.

15. In some African countries environmental legal standards and obligations are not directly implemented and enforced on customary land.

- Traditional authorities are regarded as custodians of the natural resources, including land and soils.
- Examples however demonstrate that local communities due to their internal priorities are not necessarily the best protectors of soils.
- It was recommended that environmental law provisions should be enforced on customary land.

16. A particular need was seen to establish effective policies and laws on the use of fertilizers.
17. The intervention clause of the German Nature Conservation Act was discussed as a model for implementing the Land Degradation Neutrality-concept, also in African legislation. The German Intervention clause requires three steps of decision making: (1) prevent/avoid, (2) physically offset; (3) assessing and weighing the competing interests, plus mandatory monetary compensation. This one rule captures the LDN concept and also corresponds to the UNCCD's "LDN response hierarchy".

- As a precondition for an effective use of the German Intervention clause, data are required for the determination of the baseline and the assessment of degradation and restoration processes.
- Moreover, indicators must not be too technical.

18. Restoration/offsetting requirements must not be interpreted as a "license to degrade".
19. The German example emphasizes that planning instruments are required for LDN implementation, in particular to enforce the neutrality requirement.
20. Standards and requirements for sustainable soil management must be site-specific taking into account the different settings in African countries.

21. Actual impediments in many African countries, like lacking capacities which cause insufficient and inappropriate enforcement, must be overcome.

Relevance of Land Rights/Tenure in Africa to Sustainable Soil Management

22. Clarity on land rights must be seen as a prerequisite for sustainable soil management in Africa. This is also important for medium or even long-term investments.
23. The land right systems are very different in African states. Thus one fits all-approach does not exist. Usually the following three land types are known: state land, private land and customary/tribal land.

  The terms customary land and tribal land could be used interchangeable.

24. Land tenure security (in form of western land titling/registration) is often low in Africa. In Uganda only 10% of the land is formally registered. In Kenya 67% of the land is customary in nature.
25. There was agreement that an effective land titling system is the key for land security in particular as the economic interest in land is constantly increasing, including for foreign investors.

  • A need was seen to clarify the requirements of customary land in the various communities. Cultural aspects have to be taken into account as the concept of customary land varies amongst the various communities.
  • It is to be considered that according to customs and traditions land could carry more value than the pure substance, inter alia religious or emotional values.
  • The procedure of land titling should involve "local communities" as stakeholders.
  • The procedure should be simple, cost-effective and inclusive.

    – The registration process in Uganda seems to be very time consuming and demanding.
    – The registration process in Kenya establishes a deadline which seems to contradict the concept of inclusiveness.

26. Access to land and land rights could raise gender issues.
27. "Land grabbing" is seen as a major issue in many African countries. "Land grabbing" was understood as being an "unfair" land acquisition. "Unfair land acquisition" could be both legal and illegal.
28. Foreign investment is commonly regarded as important and helpful if it is performed in a fair manner.
29. A need was seen for regulatory concepts and additional actions to deal with unfair land acquisition. Two situations were distinguished:

  • Type one: unfair use of unclear land rights.
  • Type two: unfair use of corruption potentially on all levels (state level, regional governance or municipalities)

30. Measures were discussed to deal with the two types of unfair land acquisition.

31. There was consensus that an effective land titling system is very important to allow for clarity on land rights. The procedure of land titling needs to be transparent and participatory in order to avoid corruption.

32. Two additional measures were discussed:

    • The foreign investor has to bear the burden of proof when it comes to who actually owns a certain piece of land.
    • The country of origin of the investor should be responsible for the foreign investor to act in a fair manner.

The Need for an African Soil Convention

33. Sem Shikongo (from the Ministry of Environment of Namibia) in his "words of welcome" in the first volume of the "International Yearbook of Soil Law and Policy" raised the question whether an international Soil Regulation for Africa would be a reasonable approach.

34. The 1968 African Convention on the Conservation of Nature and Natural Resources sets out general obligations. It contains specific provisions on soils, including to establish a land-use plan and thereby promoting sustainable land use and inter alia the prevention of soil erosion. The 1968 Convention however lacks in the establishment of an institutional framework to steer implementation and compliance.

35. In 2003 a revised African Convention was signed by 41 states. It also sets out soil specific regulations. In particular, it established a secretariat and a conference of Parties. The revised African Convention got into force in 2016 after attaining the 15th ratification, and in 2017, has attained the 16th ratification from Liberia.

36. Several African States have included soil related actions in their "nationally determined contributions" (NDC) under the 2016 Paris Agreement (e.g. Algeria, Kenya, Botswana, Ghana).

37. REED+ also is a method to address soil issues which are relevant at national level by internationally stimulated actions.

38. Advantages of a specific African soil instrument could be an improved coordination, joint actions and technical cooperation in particular with regard to the implementation of LDN (e.g. indicator development).

39. Three possible options to address African soil needs at the international level:

    • An African specific protocol under UNCCD

       – Problems: There is no enabling clause under UNCCD for protocols and the management would remain on international level.

    • Ratification of revised African Convention
    • Mainstreaming existing international soil provisions by stressing the African perspective

40. Whether a regional environmental convention for Africa is reasonable must be discussed further as for example the ratification and implementation of the revised African Convention was very slow and even now, the number of current ratification (16 out of 54 signatories) limits its actual output.

41. It seems to be particularly promising to use the NDC as a means to promote soil related actions in Africa. Drought, water scarcity and food security are major challenges which all are based on soils. Soils are moreover of eminent importance for mitigating climate change. This is more so because, while under the Paris Agreement, NDCs are voluntary for developing countries, the submission of NDC targets by each country amounts to an international legal commitment. This is, however, subject to provision of technical and financial support through the Paris Agreement. However, there is need for African countries, when addressing sustainable land and soil management, to also focus on the intersect with contiguous neighbors in order to address land degradation across international borders, using the climate change legal framework.

Future of International Soil Regulations

42. Existing international soil regulation is very fragmented. Although UNFCCC, UNCCD and CBD all deal with soil issues, no coherent concept exists. Based on this assumption it was asked whether and what additional international soil related provisions may be required or useful.

43. The need of an international soil regulation has been particularly promoted by IUCN, providing several groundbreaking publications and draft international instruments. The discussion has intensified since 2015 by the successive "Global Soil Weeks", having taken place in Berlin, and the newly agreed objective of a "land degradation neutral world".

44. Although soils are locally bound they carry various ecological services which could have transboundary effects, such as being essential for food security, climate change and biodiversity. Although not agreed in a legally binding manner, the sustainable management of soils is de facto as a common concern of humankind.

45. Technically speaking an international instrument on soil issues could be established as a stand-alone treaty, as protocol to either the CBD, UNFCCC or UNCCD, or as a non-binding instrument.

46. The necessary contents of such an international instrument must be further discussed.

   - The draft soil protocol lists, *inter alia* principles, rights and obligations of states, technical cooperation, organizational structure and dispute settlement.
   - There was a proposal to design the soil related obligations driver and thread specific.
   - It was discussed whether a compensation obligation of the Industry states for the long-lasting negative effects of colonialism should be included.
   - Further, measures for extra-territorial use of land were seen as necessary.

47. The LDN target setting program of UNCCD was introduced as an important tool to promote LDN. 134 states are engaged in the program. First, experience needs to be evaluated in order to improve the conceptual approaches.

48. The need of additional international soil related provisions was challenged because the political will for an additional international treaty is lacking, the negotiation of such provisions would be too time-consuming, the endeavor

would distract from the key task to implement LDN and existing norms are expected to be solely duplicated.

49. Several other proposals were put forward as first steps:

- Use of existing provisions in UNCCD, UNFCCC and CBD to promote soil issues.
- Establishment of a soil framework treaty which is based on the exiting international regulation: the new treaty should establish the links amongst the existing provisions.
- Development of model soil legislation for the national and local level
- Awareness raising and scientific information dissemination

Outlook and Next Steps

50. The Kampala Workshop was seen as very timely and constructive, but as a first step only. The Workshop was successful in raising awareness for the "soil challenge". However, further engagement was seen as crucial. As next steps and measures the following were recommended:

51. Platforms for discussions were seen as very important. Such platforms could also be used to establish networks for cooperation. The "International Yearbook of Soil Law and Policy" published by SPRINGER was highlighted as such a platform.

52. The need to outreach to policy makers was stressed. One option for an Africa wide policy debate would be to engage the Environment Committee of the African Parliament in the debate.

53. Strengthening the cooperation with natural science was also emphasized. The African Soil Partnership could provide the required scientific input. In this context the cooperation between African universities was mentioned. Funding for the establishment of such cooperation would be required.

54. Further analysis of current regulatory approaches in African countries seems to be needed. Law comparison was seen as a helpful tool to identify gaps and loopholes in existing legislation and to develop appropriate concepts for the implementation of the LDN objective.

55. Models from other—even non-African—legislation like the German intervention clause should be considered.

56. Concepts for solving the land rights impediments for sustainable soil management were seen as an essential task.

57. A workshop on soil issues in Africa in the near future which takes an interdisciplinary perspective was proposed. The workshop should provide input from science and legal experts in order to further specify models for the implementation of the LDN objective.

58. Cooperation with Technical bodies was considered to be of great value for the funding of the various actions.

59. Policies must be established to ensure a better transfer of scientific knowledge on sustainable soil management to the daily routines of African farmers.

60. Soil awareness and soil knowledge are key factors. To this end the curricula of schools and universities should include soil aspects.

Printed by Printforce, the Netherlands